VOLUME FOUR HUNDRED AND THIRTY-THREE

METHODS IN ENZYMOLOGY

Lipidomics and Bioactive Lipids: Specialized Analytical Methods and Lipids in Disease

METHODS IN ENZYMOLOGY

Editors-in-Chief

JOHN N. ABELSON AND MELVIN I. SIMON

Division of Biology
California Institute of Technology
Pasadena, California

Founding Editors

SIDNEY P. COLOWICK AND NATHAN O. KAPLAN

VOLUME FOUR HUNDRED AND THIRTY-THREE

METHODS IN
ENZYMOLOGY

Lipidomics and Bioactive Lipids: Specialized Analytical Methods and Lipids in Disease

EDITED BY

H. ALEX BROWN
Departments of Pharmacology and Chemistry
Vanderbilt University School of Medicine
Nashville, Tennessee

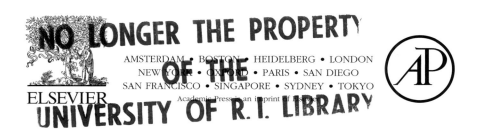

AMSTERDAM • BOSTON • HEIDELBERG • LONDON
NEW YORK • OXFORD • PARIS • SAN DIEGO
SAN FRANCISCO • SINGAPORE • SYDNEY • TOKYO
Academic Press is an imprint of Elsevier

ELSEVIER

Academic Press is an imprint of Elsevier
525 B Street, Suite 1900, San Diego, California 92101-4495, USA
84 Theobald's Road, London WC1X 8RR, UK

This book is printed on acid-free paper.

For information on all Elsevier Academic Press publications
visit our Web site at www.books.elsevier.com

ISBN: 978-0-12-373966-7

PRINTED IN THE UNITED STATES OF AMERICA
07   08   09   10   9   8   7   6   5   4   3   2   1

CONTENTS

Contributors

Kyle O. Arneson
Division of Clinical Pharmacology, Vanderbilt University School of Medicine, Nashville, Tennessee

Ian A. Blair
Centers for Cancer Pharmacology and Excellence in Environmental Toxicology, University of Pennsylvania, Philadelphia, Pennsylvania

William E. Boeglin
Division of Clinical Pharmacology, Department of Pharmacology, Vanderbilt University School of Medicine, Nashville, Tennessee

Alan R. Brash
Division of Clinical Pharmacology, Department of Pharmacology, Vanderbilt University School of Medicine, Nashville, Tennessee

Joshua D. Brooks
Division of Clinical Pharmacology, Vanderbilt University School of Medicine, Nashville, Tennessee

Raymond N. DuBois
Departments of Medicine, Cancer Biology, Cell and Developmental Biology, Vanderbilt University Medical Center, and Vanderbilt-Ingram Cancer Center, Nashville, Tennessee

Edward Felix
Department of Experimental Therapeutics, University of Texas, M. D. Anderson Cancer Center, Houston, Texas

Garret A. FitzGerald
Institute for Translational Medicine and Therapeutics, School of Medicine, University of Pennsylvania, Philadelphia, Pennsylvania

Teresa A. Garrett
Department of Biochemistry, Duke University Medical Center, Durham, North Carolina

Andrew K. Godwin
Department of Medical Oncology, Fox Chase Cancer Center, Philadelphia, Pennsylvania

Richard W. Gross

Division of Bioorganic Chemistry and Molecular Pharmacology, and Department of Internal Medicine, Washington University School of Medicine, and Department of Molecular Biology and Pharmacology, Washington University School of Medicine, and Department of Chemistry, Washington University, St. Louis, Missouri

Xianlin Han

Division of Bioorganic Chemistry and Molecular Pharmacology, and Department of Internal Medicine, Washington University School of Medicine, St. Louis, Missouri

Ronny Herzog

MPI of Molecular Cell Biology and Genetics, Dresden, Germany

Seon Hwa Lee

Centers for Cancer Pharmacology and Excellence in Environmental Toxicology, University of Pennsylvania, Philadelphia, Pennsylvania

Philip J. Kingsley

Departments of Biochemistry, Chemistry, and Pharmacology, Vanderbilt Institute of Chemical Biology, Center in Molecular Toxicology, Vanderbilt-Ingram Cancer Center, and Vanderbilt University School of Medicine, Nashville, Tennessee

Reza Kordestani

Department of Biochemistry, Duke University Medical Center, Durham, North Carolina

John A. Lawson

Institute for Translational Medicine and Therapeutics, School of Medicine, University of Pennsylvania, Philadelphia, Pennsylvania

Gerhard Liebisch

Institute of Clinical Chemistry and Laboratory Medicine, University of Regensburg, Regensburg, Germany

Shuying Liu

Department of Systems Biology, University of Texas, M. D. Anderson Cancer Center, Houston, Texas

Lawrence J. Marnett

Departments of Biochemistry, Chemistry, and Pharmacology, Vanderbilt Institute of Chemical Biology, Center in Molecular Toxicology, Vanderbilt-Ingram Cancer Center, and Vanderbilt University School of Medicine, Nashville, Tennessee

Gordon Mills

Department of Systems Biology, University of Texas, M. D. Anderson Cancer Center, Houston, Texas

Ginger L. Milne
Division of Clinical Pharmacology, Vanderbilt University School of Medicine, Nashville, Tennessee

Jason D. Morrow
Division of Clinical Pharmacology, Vanderbilt University School of Medicine, Nashville, Tennessee

Mandi Murph
Department of Systems Biology, University of Texas, M. D. Anderson Cancer Center, Houston, Texas

Robert Newman
Department of Experimental Therapeutics, University of Texas, M. D. Anderson Cancer Center, Houston, Texas

Jihai Pang
Department of Experimental Therapeutics, University of Texas, M. D. Anderson Cancer Center, Houston, Texas

Ned A. Porter
Departments of Chemistry and Medicine, Division of Clinical Pharmacology, Center in Molecular Toxicology, Vanderbilt Institute of Chemical Biology, Vanderbilt University, Nashville, Tennessee

Christian R. H. Raetz
Department of Biochemistry, Duke University Medical Center, Durham, North Carolina

L. Jackson Roberts, II
Division of Clinical Pharmacology, Vanderbilt University School of Medicine, Nashville, Tennessee

Stephanie Sanchez
Division of Clinical Pharmacology, Vanderbilt University School of Medicine, Nashville, Tennessee

Gerd Schmitz
Institute of Clinical Chemistry and Laboratory Medicine, University of Regensburg, Regensburg, Germany

Claus Schneider
Division of Clinical Pharmacology, Department of Pharmacology, Vanderbilt University School of Medicine, Nashville, Tennessee

Dominik Schwudke
MPI of Molecular Cell Biology and Genetics, Dresden, Germany

Andrej Shevchenko
MPI of Molecular Cell Biology and Genetics, Dresden, Germany

Wen-Liang Song
Institute for Translational Medicine and Therapeutics, School of Medicine, University of Pennsylvania, Philadelphia, Pennsylvania

Tamotsu Tanaka
Department of Applied Biological Science, Fukuyama University, Fukuyama, Japan

Rosanne Trost
Department of Systems Biology, University of Texas, M. D. Anderson Cancer Center, Houston, Texas

Dingzhi Wang
Department of Medicine, Vanderbilt University Medical Center, Nashville, Tennessee

Miao Wang
Institute for Translational Medicine and Therapeutics, School of Medicine, University of Pennsylvania, Philadelphia, Pennsylvania

Huiyong Yin
Departments of Chemistry and Medicine, Division of Clinical Pharmacology, Center in Molecular Toxicology, Vanderbilt Institute of Chemical Biology, Vanderbilt University, Nashville, Tennessee

Zheyong Yu
Division of Clinical Pharmacology, Department of Pharmacology, Vanderbilt University School of Medicine, Nashville, Tennessee

Yuxiang Zheng
Division of Clinical Pharmacology, Department of Pharmacology, Vanderbilt University School of Medicine, Nashville, Tennessee

Helen Zou
Institute for Translational Medicine and Therapeutics, School of Medicine, University of Pennsylvania, Philadelphia, Pennsylvania

PREFACE

Lipid metabolism and cellular signaling are highly integrated processes that regulate cell growth, proliferation, and survival. Lipids have essential roles in cellular functions, including determinants of membrane structure, serving as docking sites for cytosolic proteins and allosteric modulators. Abnormalities in lipid composition have established roles in human diseases, including diabetes, coronary disease, obesity, neurodegenerative diseases, and cancer. In the post-genomic era, we look at epigenetic factors and metabolomic biomarkers to better understand the molecular mechanisms of complex cellular processes and realize the benefits of personalized medicine.

Recent advances in lipid profiling and quantitative analysis provide an opportunity to define new roles of lipids in complex biological functions. Lipidomics was developed to be a systems biology approach to better understand contextual changes in lipid composition within an organelle, cell, or tissue as a result of challenge, stress, or metabolism. It provides an approach for determining precursor–product relationships as well as ordering the temporal and spatial events that constitute vital processes. This volume of *Methods in Enzymology* is one of a three-volume set on *Lipidomics and Bioactive Lipids* designed to provide state-of-the-art techniques in profiling and quantification of lipids using mass spectrometry and other analytical techniques used to determine the roles of lipids in cell function and disease. The first volume (432), *Mass-Spectrometry–Based Lipid Analysis,* provides current techniques to profile lipids using qualitative and quantitative approaches. The cell liposome is composed of thousands of molecular species of lipids; thus, generating a detailed description of the membrane composition presents both analytical and bioinformatic challenges. This volume includes the methodologies developed by the National Institute of General Medicine large-scale collaborative initiative, LIPID MAPS (www.lipidmaps.org), as well as an overview of international lipidomics projects. The second volume (433), *Specialized Analytical Methods and Lipids in Disease,* presents applications of lipid analysis to understanding disease processes, in addition to describing more specialized analytical approaches. The third volume (434), *Lipids and Cell Signaling,* is a series of chapters focused on lipid-signaling molecules and enzymes.

The goal of these volumes is to provide a guide to techniques used in profiling and quantification of cellular lipids with an emphasis on lipid signaling pathways. Many of the leaders in the emerging field of lipidomics have contributed to these volumes, and I am grateful for their comments in shaping the content. I hope that this guide will satisfy the needs of students who are interested in lipid structure and function as well as experienced researchers. It must be noted that many of the solvents, reagents, and instrumentation described in these chapters have the potential to be harmful to health. Readers should consult material safety data sheets, follow instrument instructions, and be properly trained in laboratory procedures before attempting any of the methods described.

H. Alex Brown

METHODS IN ENZYMOLOGY

LIQUID CHROMATOGRAPHY MASS SPECTROMETRY FOR QUANTIFYING PLASMA LYSOPHOSPHOLIPIDS: POTENTIAL BIOMARKERS FOR CANCER DIAGNOSIS

Mandi Murph,* Tamotsu Tanaka,[‡] Jihai Pang,[†] Edward Felix,[†] Shuying Liu,* Rosanne Trost,* Andrew K. Godwin,[§] Robert Newman,[†] *and* Gordon Mills*

Contents

Abstract

Cancer is a complex disease with many genetic and epigenetic aberrations that result in development of tumorigenic phenotypes. While many factors contribute

* Department of Systems Biology, University of Texas, M. D. Anderson Cancer Center, Houston, Texas
[†] Department of Experimental Therapeutics, University of Texas, M. D. Anderson Cancer Center, Houston, Texas
[‡] Department of Applied Biological Science, Fukuyama University, Fukuyama, Japan
[§] Department of Medical Oncology, Fox Chase Cancer Center, Philadelphia, Pennsylvania

Methods in Enzymology, Volume 433
ISSN 0076-6879, DOI: 10.1016/S0076-6879(07)33001-2

to the etiology of cancer, emerging data implicate lysophospholipids acting through specific cell-surface, and potentially intracellular, receptors in acquiring the transformed phenotype propagated during disease. Lysophospholipids bind to and activate specific cell-surface G protein–coupled receptors (GPCRs) that initiate cell growth, proliferation, and survival pathways, and show altered expression in cancer cells. In addition, a number of enzymes that increase lysophospholipid production are elevated in particular cell lineages and cancer patients' cells, whereas in a subset of patients, the enzymes degrading lysophospholipids are decreased. Thus, ideal conditions are established to increase lysophospholipids in the tumor microenvironment. Indeed, ascites from ovarian cancer patients, which reflects both the tumor environment and a tumor-conditioned media, exhibits markedly elevated levels of specific lysophospholipids as well as one of the enzymes involved in production of lysophospholipids: autotaxin (ATX). The potential sources of lysophospholipids in the tumor microenvironment include tumor cells and stroma, such as mesothelial cells, as well as inflammatory cells and platelets activated by the proinflammatory tumor environment. If lysophospholipids diffuse from the tumor microenvironment into the bloodstream and persist, they have the potential to serve as early diagnostic markers as well as potential monitors of tumor response to therapy. Many scientific and technical challenges need to be resolved to determine whether lysophospholipids or the enzymes producing lysophospholipids alone or in combination with other markers have the potential to contribute to early diagnosis.

Breast cancer is the most frequently diagnosed cancer among women. Mammography is associated with morbidity and has a high false positive and false negative rate. Thus, there is a critical need for biomarkers that can contribute to reduced false positive and false negative diagnoses, and to identify, stage, and/or predict prognosis of this disease to improve patient management. Here we describe a technical approach that can be applied to human blood plasma to measure the concentration of growth factor–like lysophospholipids contained in circulation. Using liquid chromatography mass spectrometry (LC/MS/MS), we quantified the amount of lysophosphatidic acid (16:0, 18:0, 18:1, 18:2, and 20:4), lysophosphatidylinositol (18:0), lysophosphatidylserine (18:1), lysophosphatidylcholine (16:0, 18:0, 18:1, 18:2, and 20:4), sphingosine-1-phosphate, and sphingosylphosphorylcholine species from human female plasma samples with malignant, benign, or no breast tumor present. Other methods described here include handling patient blood samples, lipid extraction, and factors that affect lysophospholipid production and loss during sample handling.

1. INTRODUCTION

An estimated 200,000 new cases of female breast cancer were diagnosed in the United States in 2006, making breast the most frequently occurring cancer in women (American Cancer Society, 2006). Since the early 1990s, long-term survival rates for breast cancer have increased consistently every year as routine screening increases and new therapies become available; however, recurrence and metastases continue to be a problem among survivors, resulting in marked morbidity and mortality, with breast cancer being the second most common cause of death from cancer. For many women with localized disease, the risk of recurrence is low, but the challenge is to separate this group from those who will likely have a recurrence. The ability to distinguish these groups would (1) reduce the amount of over-treatment and long-term toxicity in patients who are cured by local treatment, and (2) dictate more aggressive treatment and monitoring of those patients who are more likely to experience recurrence. The diversity of individual breast cancer subtypes is incredibly high, and thus biomarkers are needed to distinguish between these patient populations.

Biomarkers can be almost any detectable protein, DNA, RNA, or diagnostic molecule found in human fluids that indicates the risk or presence of disease. The prostate-specific antigen (PSA) is a routinely used screening biomarker designed for the early detection of prostate cancer. Should PSA or any other biomarker indicate abnormality, further diagnostic tests are performed to confirm the disease state. Indeed, PSA and other biomarkers should not be considered a "screening" test, but rather as a "prescreen" designed to determine which individuals are at high risk and warrant further investigation. Mammography plays a similar role, identifying a population of patients who require additional investigation, with biopsy and pathology being the "second-stage" screen for high risk individuals. Multiple studies, albeit controversial, have suggested that lysophospholipids, which are abundant in human blood, should be considered and evaluated for their potential as biomarkers. A phospholipid is a lipid molecule that contains glycerol and fatty acyl chains, and has a phosphate group attached to it, whereas a lysophospholipid is a phospholipid that is missing one of its fatty acyl chains. In addition to lysophospholipids, sphingolipids—which have a sphingosine backbone—clearly play a role in tumor development and progression (Spiegel and Milstien, 2003; Visentin et al., 2006).

Lysophospholipids are well-known activators of proliferation and growth in breast cancer cells (Xu et al., 1995) and natural enhancers of metastatic potential through initiating migration (Sliva et al., 2000). Indeed, progression to bone metastasis, one of the major causes of morbidity from breast cancer, was recently attributed to the action of the endogenous lysophospholipid ligand, lysophosphatidic acid (LPA), and the overexpression of the lysophospholipid receptor LPA_1, in the tumor microenvironment (Boucharaba et al., 2006). Increased expression of a homologous LPA G protein–coupled receptor (GPCR), LPA_2, is present in approximately 57% of breast carcinoma cases examined compared with normal epithelial cells (Kitayama et al., 2004). Taken together, emerging evidence suggests that bioactive lysophospholipids like LPA and its receptors may contribute to breast cancer progression.

Previous research on ovarian cancer has demonstrated that the levels of LPA are likely elevated in the ascites and malignant effusions of patients (Baker *et al.*, 2002; Xu *et al.*, 1998). A similar lysophospholipid, lysophosphatidylinositol (LPI), was reported to have a comparatively high level in the plasma of ovarian cancer patients (Xiao *et al.*, 2000). Likewise, the LPA precursor lysophospholipid, lysophosphatidylcholine (LPC), is elevated in the plasma (Okita *et al.*, 1997) and ascitic fluid of these patients (Xiao *et al.*, 2001). Other groups have shown an altered fatty acid composition of LPC in the plasma and serum of ovarian cancer patients (Sutphen *et al.*, 2004; Tokumura *et al.*, 2002b). Recent evidence suggests that lysophosphatidylserine (LPS) is a ligand for GPR34, a previously orphaned GPCR that is highly expressed on mast cells where LPS can induce activation (Sugo *et al.*, 2006). These data suggest that lysophospholipids may be viable biomarkers in ovarian and possibly other cancers. However, this area remains controversial because of challenges with production and loss of LPA during handling as well as difficulties in measurement, particularly of isoforms of LPA.

In addition to the aforementioned LPA, LPI, LPC, and LPS, other bioactive lysophospholipids and related sphingolipids contribute to the progression of cancer. For example, sphingosine 1-phosphate (S1P) is a proven inducer of angiogenesis (Lee *et al.*, 1999) and vasculogenesis (Argraves *et al.*, 2004), and contributes to growth and invasion in a variety of cell types (Visentin *et al.*, 2006). S1P can be targeted by immunoneutralizing antibodies, inhibitors of sphingosine kinase, and inhibitors of S1P receptors. A novel immunoneutralizing antibody (Sphingomab™; Lpath, Inc, San Diego, CA), and the FTY720, a pro-drug receptor inhibitor, have demonstrated activity in preclinical animal models (Milstien and Spiegel, 2006; Visentin *et al.*, 2006). Theragnostics, diagnostic tests related to determining levels of drug targets and identification of individuals likely to respond, are clearly needed to select patients for therapy with these agents. The Sphingomab antibody demonstrates marked specificity for S1P, which exists as a single isoform in patients, providing the opportunity for analysis of S1P levels in blood or tumors using affinity binding assays. Although high-affinity monoclonal antibodies for LPA have been developed, they do not distinguish among the various isoforms of LPA. Thus, while they may be useful in determining total LPA levels, they will need to be combined with other approaches to determine the levels of particular LPA isoforms in blood or tumors.

S1P is produced by adding a phosphate group to sphingosine through the action of sphingosine kinase 1 and 2 (SphK1 and SphK2). Increased expression of SphK1 promotes tumorigenesis and the progression of breast cancer through regulation of S1P production, growth, and survival (Nava *et al.*, 2002). In addition, SphK1 is required for EGF-directed motility in MCF-7 breast cancer cells, and its downregulation may allow enhanced sensitivity to chemotherapy (Sarkar *et al.*, 2005). In other cell types, such as rat epithelial cells, SphK1 induces COX-2 expression and contributes to colon carcinogenesis (Kawamori *et al.*, 2006), and its expression is required for cell proliferation of intestinal tumors (Kohno *et al.*, 2006). Higher than normal levels of SphK1 expression were observed with lung cancer (Johnson *et al.*, 2005), oncogenic transformation in erythroleukemia (Le Scolan *et al.*, 2005), and also correlate with short-term patient survival with glioblastoma multiforme (Van Brocklyn *et al.*, 2005).

Autotaxin (ATX), originally identified as a motility and metastases promoting factor released by melanoma cells (Nam *et al.*, 2001; Stracke *et al.*, 1992) is the primary enzyme regulating the production of LPA from LPC (Tokumura *et al.*, 2002a; Umezu-Goto *et al.*, 2002). Indeed, heterozygous ATX knockout mice have circulating LPA levels of about one-half of wild-type littermates confirming the primacy of ATX in regulating LPA levels (van Meeteren *et al.*, 2006). ATX is exquisitely regulated by autoinhibition by LPA, resulting in tight control of circulating LPA levels despite a high level of the LPC precursor in plasma (van Meeteren *et al.*, 2005). Strikingly, cyclic PA analogues that function as inhibitors of ATX are potent inhibitors of metastases in a number of tumor models establishing LPA as a key regulator of metastases (Baker *et al.*, 2007). Aberrant overexpression of ATX may contribute to tumor progression with elevated levels seen in glioblastoma multiforme (Kishi *et al.*, 2006), human hepatocellular carcinoma (Zhang *et al.*, 1999), undifferentiated anaplastic thyroid carcinoma cell lines (UTC) (Kehlen *et al.*, 2004), poorly differentiated non–small-cell lung carcinomas (Yang *et al.*, 1999), and breast cancer cells where ATX is associated with invasiveness (Yang *et al.*, 2002). Furthermore, ATX transgenic mice demonstrate increased incidence of invasive and metastatic cancers. ATX is released from tumor cells by an as yet unclear mechanism. In addition, vesicles released from tumor cells as a consequence of PLD activity are accessible to phospholipase A1 and A2, potentially increasing the amount of LPC available to ATX. In addition to ATX, which increases LPA production and is abnormally expressed in some cancers, the lipid phosphate phosphatases (LPP1-3) are reduced in a number of cancer cell lineages and patient tumors (Tanyi *et al.*, 2003a,b). Further, reintroduction of LPPs into cancer cells decreases growth and survival *in vitro* and *in vivo* (Tanyi *et al.*, 2003a,b).

Taken together, there are marked aberrations in the production and action of lysophospholipids in particular cancer patients. Increased local production or production of lysophospholipids in the bloodstream likely contributes to the initiation and progression of cancers including ovarian, breast, and other tumor lineages. Because many of the enzymes that metabolize lysophospholipids are aberrantly expressed in cancer, lysophospholipids in plasma are potential biomarkers for early detection and for distinguishing malignant from benign and normal conditions. In this chapter, we describe methods for handling patient blood samples, lipid extraction, and LC/MS/MS analysis to quantify lysophospholipids in plasma.

2. Procedure

2.1. Materials for liquid chromatography/mass spectrometry/ mass spectrometry

1-acyl LPA species with 16:0, 17:0, 18:0, 18:1, 20:4; 1-acyl LPC species with 16:0, 17:0, 18:0, 18:1; 1-acyl LPI with 18:0; 1-acyl LPS with 18:1; S1P; and SPC were purchased from Avanti Polar Lipids (Alabaster, AL). 1-acyl LPA with 18:2 was purchased from Echelon Biosciences, Inc. (Salt Lake City, UT). 2-acyl LPC species with 18:2 and 20:4 were chemically synthesized by acid hydrolysis of 1-alkenyl-2–18:2 and 1-alkenyl-2–20:4 phosphatidylcholine (Satouchi *et al.*, 1994). These alkenyl–acyl phosphatidylcholine species were synthesized by condensation of fatty acyl anhydride, and 1-alkenyl-2-lysophosphatidylcholine prepared from bovine heart, also from Avanti (Satouchi *et al.*, 1994). To confirm the results, chemically synthesized LPC species were verified by mass spectrometry analysis. The synthesized 2-acyl LPC species were used as standards for LC/MS/MS analysis within 1 week after preparation.

2.2. Collection of patient samples

Blood samples were obtained at the University of Texas M. D. Anderson Cancer Center and Fox Chase Cancer Center (Philadelphia, PA). All blood samples were collected using acid citrate dextrose (ACD) as the anticoagulant. In order to minimize effects of the bioactive enzymes ATX and lecithin–cholesterol acyl transferase (LCAT) on lysophospholipids levels, plasma samples were processed and frozen within 4 h after obtaining them from patients and healthy controls. ACD tubes containing samples were allowed to settle for 10 min at room temperature prior to centrifugation at $1300 \times g$ for 7 to 10 min. Plasma samples were then aliquoted into cryovials and frozen immediately at $-70°$.

2.3. Extraction of lipids

Frozen patient plasma was thawed at $4°$, transferred to glass test tubes, and acidified using 40 μl of 1 N citric acid (Sigma Aldrich, St. Louis, MO) per 0.5 ml of plasma. Phosphate buffered saline (PBS) at volume equal to plasma and 3 ml of 1-butanol were added and vortexed. Internal standard was added to each sample (0.1 nmol of 17:0 LPA and 10 nmol of 17:0 LPC), and samples were vortexed for 1 h at room temperature. After vortexing, the samples were centrifuged at 4000 rpm for 10 min. The upper phase of the sample was transferred to a glass collection tube, and 1.5 ml of water-saturated butanol was added to the original sample tube's lower phase.

These samples were then vortexed again for 30 min prior to centrifugation at 4000 rpm for 10 min. The upper phase was removed again and combined with the first upper phase in the glass collection tube. The organic solvent was evaporated under a nitrogen stream at 40°. All evaporated samples were kept at −20° until analysis.

2.4. Analysis of incubated plasma

Approximately 0.5 ml of EDTA or heparin plasma were incubated for 24 h in glass test tubes at 37° with continual shaking (230 rpm). Lipids were extracted as described above after incubation and fractionated with TLC using the solvent system (chloroform:MeOH:20% NH_3=60:35:8, v/v). The bands corresponding to LPC and LPA were scraped off the plate and extracted. The amounts of LPA and LPC were then determined by the Bartlett method (1958).

3. Liquid Chromatography/Mass Spectrometry/Mass Spectrometry

To analyze LPA species, we first mixed the dried lipid extract from plasma with 0.5 ml of methanol, vortexed the samples for 2 min and then centrifuged them at 5000 rpm at 2 min. The lysophospholipids of interest were present in the methanol soluble fraction. This was followed by mixing 50 μl of the methanol-soluble fraction with 50 μl of the mobile phase (acetonitrile: 10 mM NH_4OCOCH_3=1:1). Ten microliters of this solution were analyzed by liquid chromatography/mass spectrometry/mass spectrometry (LC/MS/MS) using a Waters Quattro Ultima tandem mass spectrometer (Milford, MA) operated in electrospray negative-ionization mode. The LC system was an Agilent 1100 binary HPLC (Wilmington, DE). Chromatographic resolution of individual LPA species was accomplished using a linear gradient of 10 mM ammonium acetate, pH 8.5 buffer (mobile phase A), and 98: 2 acetonitrile-methanol (mobile phase B) with a Phenomenex Luna C5 5 μm, 150 × 2.0 mm analytical column (Torrance, CA). The LC gradient conditions consisted of 35% B from 0 to 0.5 min, from 35% B to 100% B at 6 min, holding at 100% until 9 min, returning to 35% B at 9.5 min, and then holding at 35% B until 15 min. The column temperature and flow rate were maintained at 45° and 0.3 ml/min throughout the analysis. The mass spectrometer source temperature was 125°, desolvation temperature was 350°, and nitrogen gas nebulization flow was 900 liters/h.

The method for detection of LPC species follows: 50 μl of methanol-soluble fraction was mixed with 50 μl of 1:1 methanol:0.05% formic acid (aq.), pH 3.0, and 15 μl of this mixture was analyzed by LC/MS/MS. The LC/MS/MS instrumentation was operated in electrospray positive-ionization mode, and chromatographic resolution of LPCs was accomplished using a linear gradient of 0.05% formic acid (aq.), pH 3.0 (mobile phase A), and 90:10 methanol:acetonitrile (v/v) (mobile phase B) with a Phenomenex Luna C18-HC, 150-×-2.0–mm analytical column (Torrance, CA). The gradient consisted of 25% B from 0 to 0.5 min, rising to 100% B at 2.0 min, and held at 100% B until 9.5 min, and then dropping to 25% B at 10 min and holding at 25% B until 15 min. Column temperature and flow rates were maintained at 65° and 0.3 ml/min throughout analysis. Source temperature, desolvation temperature, and nebulization gas flow rate was the same as with LPA settings. Specific precursor and product ion transitions used for the determination of individual LPAs and LPCs are listed in Fig. 1.2A and B.

3.1. Analysis of lysophospholipid recovery rate

The lysophospholipids were extracted from 0.5 ml of plasma with or without a "test" loading of 2 nmol of synthetic lysophospholipids (17:0 LPA, 18:0 LPI, 18:1 LPS, S1P) or 20 nmol of lysophospholipids (17:0 LPC and SPC). The lysophospholipids in the extracted sample were then analyzed by LC/MS/MS as described above. A nonextracted reference standard was also prepared by mixing 2 nmol or 20 nmol of each synthetic lysophospholipid and analyzed by LC/MS/MS as described above. Based on the relative peak area between extracted samples that had been spiked with standard lysophospholipids and nonextracted reference samples, the recovery of the lysophospholipids was calculated. Determination of the peak area for a specific lysophospholipid in a spiked sample was obtained by the subtracting the peak area of the specific lysophospholipid as measured in an extracted blank sample. This was necessary because of the high endogenous levels of LPAs and LPCs in plasma.

3.2. Standard curve calculation for analysis of plasma

LPA and LPC standard calibration curves were prepared daily to compensate for day-to-day fluctuations in electrospray ionization efficiency. Lysophospholipid calibration curves were prepared by placing each individual LPA or LPC species to be measured into a glass tube and evaporating to dryness. The dried LPA or LPC mixture was then redissolved in chloroform/methanol (2:1, v/v). Serial dilutions of this mixture are then prepared. To each dilution a fixed amount of either 0.1 nmol 17:0 LPA or 10 nmol 17:0 LPC was added as internal standard for respective analysis of LPAs or LPCs. Healthy control plasma was then added to each glass tube and vortex mixed. Sample plasma extraction was performed as described previously.

The standard calibration curve was then plotted based on molar concentration versus peak area ratio of each specific LPA or LPC to their respective internal standard. The concentration range of quantification for LPAs and LPCs were 0.5 to 20 nM and 5 to 100 pM, respectively.

3.3. Determination of correction factor for LPC isomers

The standard sn-2–18:2 LPC isomer was prepared as described above. The sn-1–18:2 LPC isomer was synthesized by phospholipase A2 treatment of 1,2 di-18:2 phosphatidylcholine (Tanaka *et al.*, 2004). Each 18:2 LPC isomer was mixed with a fixed amount of 17:0 LPC at different molar ratio (17:0LPC/18:2 LPC=1: 0.5, 1:1, or 1:5) and analyzed immediately. Standard curves were obtained by plotting the peak area ratio against molar ratio to calculate the slope. We found that sn-2–18:2 LPC is more efficiently detected than sn-1–18:2 LPC under our analytical conditions. We used sn-2–18:2 LPC as our standard curve sample and used the values determined from the slope for the correction factor in our calculations of sn-1–18:2 LPC isomers.

3.4. Statistical analysis

The statistical analysis of data was performed by 1-way analysis of variance (ANOVA) followed by Tukey's test.

4. DISCUSSION

Our study began with the careful construction of conditions before analyzing any human samples. All steps should be carried out using glass tubes and pipettes, and it is critical that the protocol followed for human blood collection is appropriate for analysis of lysophospholipids. For example, in order to accurately determine the amount of initial plasma LPA concentration, a calcium-trapping anticoagulant, such as EDTA or ACD, should be used. Both EDTA and citrate chelate calcium and decrease the ability of LPA-producing enzymes like ATX to actively form LPA during sample processing. Therefore, no additional LPA formation should occur during the handling of the sample, giving a baseline lipid measurement. As indicated in Fig. 1.1, incubation of heparinized plasma at 37° for 24 h results in a modest increase in S1P, significant increases in LPC (particularly based on basal levels of above 100 nm/ml), and at least a 15-fold increase in LPA levels. The concurrent increases in LPA and LPC indicate that the amount of LPC converted to LPA is not sufficient to overwhelm the production of LPC during incubation. It is important to note as previously demonstrated

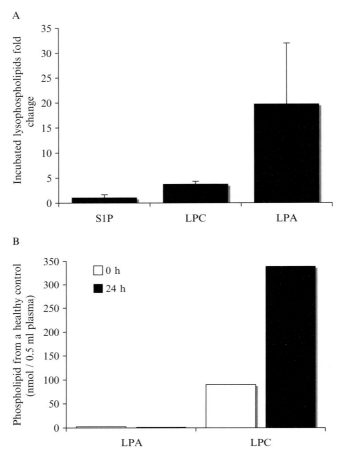

Figure 1.1   Incubating heparin-plasma (A), but not EDTA-plasma (B), increases LPA concentration. (A) Lipids from heparin-plasma samples of 10 healthy controls were extracted after incubating for 24 h at 37° with continuous shaking. S1P, LPA, and LPC fractions were isolated by TLC. The amounts of lysophospholipids were then determined by the method of Bartlett (1958). Lipids from nonincubated plasma was also analyzed to achieve the fold change. Data presented comprise the average-fold change of the samples with the standard deviation among samples. (B) Concentration of LPA and LPC from EDTA-plasma samples before (0 h) and after (24 h) incubation were analyzed. The basal levels of LPA and LPC were 2.3 and 90.5 nmol/0.5 ml plasma, respectively. LPA incubation did not result in a fold increase. LPC levels increased nearly four-fold.

by others that there is an active LPA regeneration system present during culture that results in increased LPA levels over time, which potentially explains the growth- and survival-promoting activity of fetal bovine serum in tissue culture. In contrast, in EDTA plasma, there were no detectable alterations in LPA levels, whether positive or negative. Thus, it is likely that LPA levels in ACD or EDTA plasma accurately reflect intravascular LPA levels (excluding possible sources of loss or non–calcium-dependent mechanisms of production, such as release from platelets). In contrast, neither heparin nor calcium-trapping anticoagulants blocks production of LPC (Fig. 1.1A and B). Thus, LPC levels detected in heparin or EDTA plasma likely represent a combination of post-collection generation and loss, and may not reflect levels *in vivo*. Indeed, even the short times from collection to preparation of plasma and freezing could result in markedly altered LPC levels.

Acidic lysophospholipids such as LPA and S1P are not recovered by organic solvent extraction at neutral pH; therefore, mild acidification of plasma is necessary. We used citric acid to acidify plasma samples as previously reported (Baker *et al.*, 2001), because adding high concentrations of HCl to plasma results in the artificial formation of LPA from LPC during the extraction (Ishida *et al.*, 2005). We next used 1-butanol as the organic solvent to extract the upper phase, followed by water-saturated butanol for the lower phase extraction (Baker *et al.*, 2001).

To verify conditions, we first performed extractions using standard lipids spiked into human control plasma to determine the amount of lipid recovery. In all lysophospholipids measured, the percent of lipid recovery was sufficiently high for accurate detection with our methods (Table 1.1), ranging from 60% for LPI to 93% for LPA and higher for other lipids assayed. The lipid extracts obtained from plasma were fractionated into methanol-soluble and -insoluble fractions during the final step. While considerable lipidic components are present in the methanol-insoluble fraction, these data indicate that lysophospholipids are recovered from the methanol-soluble fraction.

In addition, a standard calibration curve was calculated for each lipid measured, and an internal standard—17:0 LPA for LPA or 17:0 LPC for LPC analyses—was used to monitor sample recovery. This is important for the analytical conditions, including detection efficiency in LC/MS/MS,

Table 1.1 Percentage of lysophospholipid recovery after extraction and LC/MS/MS

Lysophospholipids	Recovery (%)
17:0 LPA	93.3 ± 0.1
LPI	59.9 ± 4.8
LPS	77.6 ± 3.1
17:0 LPC	97.0 ± 7.0
S1P	91.6 ± 16.7
SPC	96.0 ± 8.2

because it is not always the same each day and even over the course of a single day. Therefore, the standard curve of each lysophospholipid species was obtained in every analysis and used to quantify levels. Excellent linear correlations between the peak area ratio and molar ratio were achieved at relatively wide range (molar ratio: 0.5 to 40 for LPAs, and 0.5 to 10 for LPCs; Fig. 1.3).

The stability of the LC/MS/MS instrument during analysis was carefully monitored by inserting quality controls between every 10 samples. MS/MS detection of LPI is shown in Fig. 1.2A and B. The set of deprotonated molecular ion and its daughter ion for LPI (18:0) is *m/z* 599 and 283,

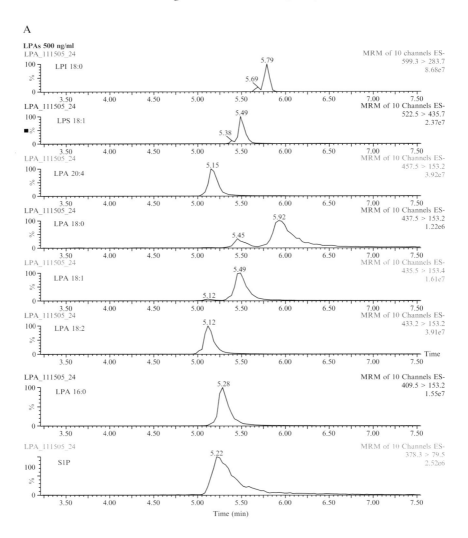

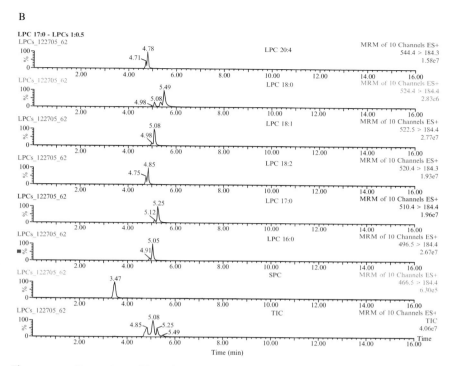

Figure 1.2 Detection of lysophospholipids at negative (A) and positive (B) mode. (A) Mixture of standard LPS (18:1), LPI (18:0), LPA (16:0, 18:0, 18:1, 18:2 and 20:4), and S1P was analyzed and detected by negative MS/MS using sets of deprotonated molecular ion [M-H]⁻ and its daughter ion as described in section 4. (B) A mixture of LPC (16:0, 18:0, 18:1, 18:2, and 20:4) and SPC were detected by positive MS/MS. The sets of protonated molecular ions [M+H]⁺ and daughter ion m/z 184 [phosphorylcholine]⁺ were used. LPC 18:2 shows two peaks, which indicate the presence of both isomers—2-acyl type (fast-migrating peak) and 1-acyl type (slow-migrating peak). (See color insert.)

respectively. Possible structure of the m/z 283 is [stearic acid -H]⁻. The set of the ions for detection of LPS (18:1) is m/z 522 and 435, respectively. Possible structure of the daughter ion is [18:1 LPA - H]⁻. Sets of deprotonated molecular ion and m/z 153 [glycerophosphate -H₂O -H]⁻ can be used for LPAs. For detection of S1P, the set of m/z 378 and m/z 79 can be used. The daughter ion is assigned to be [PO₃]⁻. Detection of LPC and SPC can be done with positive mode. The protonated molecular ion [M+H]⁺ and m/z 184 is the ion set for detection. Possible structure of m/z 184 is [phosphorylcholine]⁺.

Our studies revealed that in 16:0, 18:0, and 18:1 LPC, the most prominent isomer is 1-acyl, and in 20:4 LPC, the 2-acyl form is the most prominent isomer. This may be particularly important, because it has been suggested that different LPA receptors have a preference for the

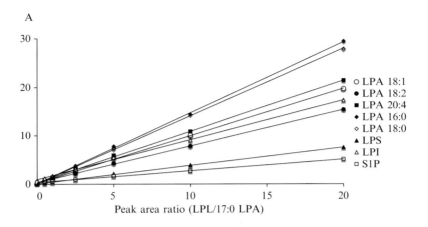

Figure 1.3 Calibration curves for lysophospholipids. (A) Standard curves for LPA 16:0 (*closed diamonds*), LPA 18:0 (*open diamonds*), LPA 20:4 (*closed squares*), LPA 18:1 (*open circles*), LPI (*open triangles*), LPA 18:2 (*closed circles*), LPS (*closed triangles*), and S1P (*open squares*). (B) Slope of the lines in A.

1- and 2-acyl forms of LPA. They may also be important for identifying the pathway involved in LPA production. In contrast, for 18:2 LPC we detected both 1-acyl and 2-acyl isomers in significant amounts. Because of their instability, these lipids are commercially unavailable and were synthesized from phosphatidylcholine on the same day, and immediately mixed with 17:0 LPC in different ratios before analysis with LC/ESI/MS. Determination of 18:2 (2-acyl type) LPC can be done using a correction factor, because we found that 2–18:2 LPC is 2.9 times more efficiently detected than the corresponding 1-acyl type when we analyze standard *sn*-1–18:2 and *sn*-2–18:2 LPC.

The positional isomers of lysophospholipids are always a problem for analysis by LC/MS because of the inability of LC/MS to resolve 1-acyl and

2-acyl isomers of LPA species. This is further complicated by the observation that the fatty acyl residue at the sn-2 position of lysophospholipids may spontaneously migrate to the sn-1 position. Although the standard LPAs used in this study are 1-acyl isomers, 2-acyl LPA would likely be present in patient samples. If distinguishing between 1-acyl and 2-acyl isomers is necessary, complete chromatographical resolution of each isomer and comparison to the appropriate standard curve may be the best approach. It should be emphasized that the approaches described herein do not have the ability to distinguish between 1- and 2-acyl forms of LPA.

We next quantified human plasma samples from women who had malignant, benign, or no breast tumor present. It is important to emphasize that the cancer and benign samples in this study were prospectively collected in a screening clinic for women with abnormal mammograms or without knowledge of the underlying diagnosis. The truly prospective collection approach removes many of the potential biases associated with retrospective sample sets. Further, the individuals running the assays were blinded as to the identity of the samples. We measured total as well as specific species of S1P, LPI, LPS, LPA, SPC, and LPC (Table 1.2 and Fig. 1.4), and found no statistically significant difference among these groups

Table 1.2 Average lysophospholipid concentration (nmol/0.5 ml plasma) in plasma of women with malignant and benign breast tumors and of healthy controls

	Cancer ($n = 26$)	Benign ($n = 27$)	Healthy ($n = 25$)
S1P	0.256 ± 0.060	0.293 ± 0.067	0.289 ± 0.087
LPS (18:1)	0.010 ± 0.005	0.012 ± 0.009	0.015 ± 0.009
LPI (18:0)	0.066 ± 0.02	0.079 ± 0.032	0.065 ± 0.022
LPA (16:0)	0.360 ± 0.076	0.376 ± 0.084	0.349 ± 0.111
LPA (18:0)	0.085 ± 0.026	0.104 ± 0.051	0.086 ± 0.038
LPA (18:1)	0.139 ± 0.031	0.145 ± 0.048	0.134 ± 0.046
LPA (18:2)	0.345 ± 0.107	0.351 ± 0.108	0.306 ± 0.092
LPA (20:4)	0.315 ± 0.129	0.332 ± 0.164	0.349 ± 0.162
Total LPA	1.244 ± 0.262	1.307 ± 0.355	1.224 ± 0.368
SPC	0.011 ± 0.009	0.008 ± 0.005	0.010 ± 0.010
LPC (1–16:0)	60.19 ± 14.91	61.09 ± 21.46	63.35 ± 22.94
LPC (1–18:0)	13.62 ± 3.59	13.86 ± 6.13	13.75 ± 5.33
LPC (1–18:1)	9.11 ± 2.64	8.52 ± 2.94	8.35 ± 3.73
LPC (1–18:2)	26.62 ± 11.54	27.14 ± 11.55	21.55 ± 7.78
LPC (2–18:2)	6.96 ± 2.72	5.66 ± 2.11	4.89 ± 2.08
LPC (total 18:2)	33.58 ± 13.22	32.80 ± 12.78	26.44 ± 7.99
LPC (2–20:4)	12.52 ± 4.83	9.27 ± 4.00	9.73 ± 5.64
Total LPC	129.02 ± 31.47	125.55 ± 35.34	121.62 ± 35.22

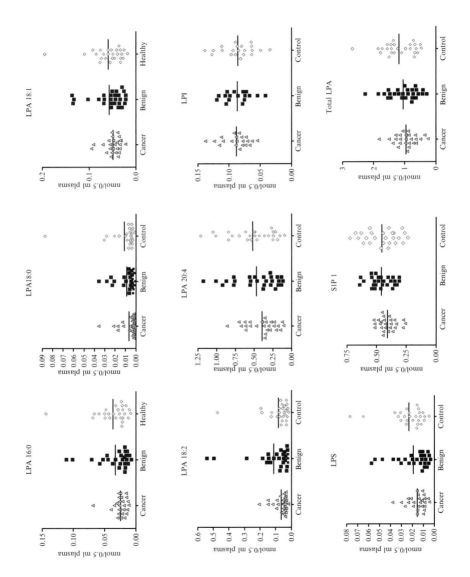

16

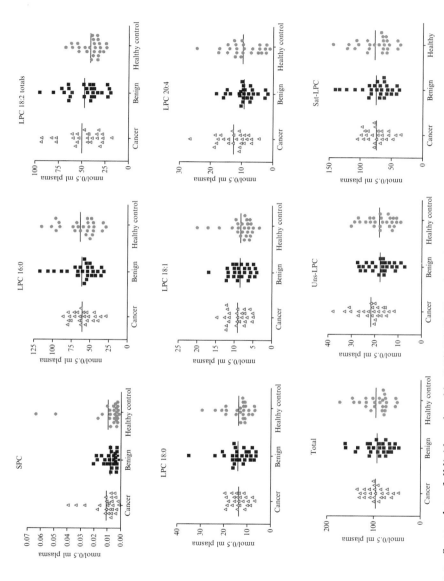

Figure 1.4 Scatterplots of all lipids analyzed by LC/MS/MS. Plots show the comparison between lipids measured in cancer patients, women with benign tumors, and healthy controls. For more details, see section 4. Results were not statistically significant. (See color insert.)

that would allow a lipid-based accurate classification of samples. Further, simple algorithms based on different forms (saturated/unsaturated), ratios, multiplications, or additions failed to identify patterns of lysophospholipids that had useful predictive values. While there were few cancer patients with levels of different isoforms of LPC outside the range found in healthy individuals, sensitivity and specificity were insufficient, particularly when compared to benign disease, to be clinically relevant. Concerns about generation and degradation of LPC isoforms during analysis increase the likelihood that these levels are not useful in screening approaches. Women with malignant cancer in this study represented all stages of breast cancer (I to IV), had breast tumor sizes ranging from 0.5 to 5 cm, and had cancer present in 0 to 10 lymph nodes. All had infiltrating ductal carcinoma present on histology staining, varying grades of differentiation, and single tumors were found in all quadrants of the breast. Only one patient had significant metastatic disease at presentation, but did not show an increase in lysophospholipids levels beyond the other groups. Thus, the samples tested represent a great diversity among breast tumors present in the general population. The negative results are likely a reflection of small, localized tumors associated with breast cancer that would be required to constantly produce an extraordinarily large volume of lysophospholipids to allow detection in general blood circulation. Based on this sample set, clinically relevant differences in levels of the LPA and LPC species characterized either do not exist or are simply too small for detection. However, it is important to emphasize that although care was taken to reduce the potential that LPA or LPC would be produced, metabolized, or lost between collection and analysis, it is impossible to eliminate this possibility. Indeed, this remains the most significant challenge to determining the usefulness of lysophospholipids in the early diagnosis of cancer.

Herein we have presented a method for quantifying lysophospholipids in plasma using liquid chromatography mass spectrometry. While this process is time consuming and requires expensive equipment, it is likely the most accurate method for detecting lysophospholipids in biological samples to date. Previously, other methods have been employed, such as thin layer chromatography coupled with gas chromatography (Shen *et al.*, 2001; Xu *et al.*, 1998), electrospray ionization mass spectrometry (Xiao *et al.*, 2000, 2001), electrospray ionization liquid-chromatography mass spectrometry (Baker *et al.*, 2001), and time-of-flight mass spectrometry using a phosphate-capture molecule (Tanaka *et al.*, 2004). Other groups continue to develop rapid methods to measure lysophospholipids from human blood and eliminate the necessity of mass spectrometry equipment through radioenzymatic detection (Saulnier-Blache *et al.*, 2000), colorimetric absorbance assays (Kishimoto *et al.*, 2002, 2003), and fluorescence assays (Alpturk *et al.*, 2006).

5. CHALLENGES TO THE DEVELOPMENT OF A CLINICALLY APPLICABLE SCREENING ASSAY

Major challenges remain in the development of a clinically applicable screening assay for any cancer lineage based on lysophospholipids. The first of these is to develop practical methods for handling samples and measuring levels of lysophospholipids in serum and plasma. The second is to determine the specificity, sensitivity, and selectivity of lysophopholipids as markers for cancer and then to determine their utility.

The development of a practical approach to analysis applicable for large patient populations is not trivial. The first approach will be to develop clinically applicable collection, handling, and storage conditions that do not alter levels of lysophospholipids through production, degradation, or absorption, to collecting and handling materials. Although there has been progress in this regard as described above and by others, there is no consensus as to whether any of the handling approaches are adequate to prevent gain or loss of lysophospholipids. The second major challenge is in prepurification of lysophospholipids from plasma or serum. Organic solvent–based technologies are not acceptable in a clinical laboratory environment unless they can be utilized in a self-contained "cassette" or "lab-on-a-chip" approach that prevents exposure of personnel to organic solvents. Affinity approaches such as the S1P and LPA antibodies produced by Lpath Inc or other affinity reagents provide a major opportunity to concentrate lysophospholipids in a useful form (Visentin *et al.*, 2006). If total levels of lysophospholipids such as S1P or LPA provide useful information, a competition-type assay may prove useful. However, since LPA and other lysophospholipids have multiple isoforms, and because knowing the amount of each isoform may be important, these antibodies will not be useful as direct assays. Nonetheless, they may prove to be effective affinity reagents for first-step purification followed by a mass-spectrometry type of assay. The only efficient approaches to assess various isoforms of lysophospholipids are based on mass spectrometry. While it is possible to adapt high-throughput assays to mass spectrometry, the assays would need to be run in centralized laboratories and costs may be too high for a screening assay. The centralized laboratory concept would add to costs and also to the potential for changes in lysophospholipid levels during handling. A number of "lab-on-a-chip" approaches could potentially be applied to lysophospholipids as well as other affinity and detection approaches. However, these approaches are in their infancy and will require additional development.

Because it is relatively easier to deal with proteins than lysophospholipids, it may be possible to develop methods to assess the levels or activity of the enzymes involved in the production and degradation of lysophospholipids.

Based on enzyme activity assays, ATX, for example, appears to have a very narrow dynamic range with limited, if any, sensitivity and specificity as a diagnostic marker in plasma or serum in diseases thus far assessed (Umezu-Goto *et al.*, 2004).

Once the technical issues are resolved, it will be necessary to determine whether there are sufficient sensitivity, specificity, and selectivity to justify use as an assay. It is important to understand the relative importance of each in developing an assay. Cancers can be divided into groups based on a number of criteria. For cancers with high prevalence, such as breast, lung, bowel, and prostate, the requirement of high specificity is not as important as for less common tumors. For example, mammography in particular patient populations has a sensitivity and specificity around 80%, and yet is considered useful. For cancers that are relatively uncommon such as ovarian or pancreatic (the fifth and sixth most prevalent cancers, about 1 in 2500 people will have these cancers at any one time), a specificity approaching 99.6 is required for a stand-alone assay to discover 1 cancer for every 10 false positive tests. For these relatively uncommon cancers, a secondary test would be required; however, the best current assays are still considered to require around 98% specificity, that is, no more than 2 false positives for every 100 tests. As an example, a single CA125 measurement has a sensitivity of approximately 70% and a specificity of approximately 90%, which is considered inadequate as a screen for ovarian cancer. For relatively rare cancers, sensitivity is not as important as specificity in determining the utility of a screen, but does contribute to concerns and risks of false assurances. Cancers can also be divided into those where there is a practical secondary screen. For example, the Pap (Papanicolaou) smear has a sensitivity approaching 100%, but a specificity of only about 50%. The low specificity is acceptable because of the availability of colposcopy and biopsy as secondary and tertiary screens. Similarly, the relatively low specificity of mammography is acceptable because of the high prevalence of the disease and to the availability of an acceptable follow-up set of studies including magnetic resonance imaging, ultrasound, and eventually biopsy. Less consideration is given to the challenges of selectivity. In terms of patient management, selectivity is comparable to differential diagnosis. A test that indicates likelihood of cancer but not location can lead to major problems and challenges particularly in the event of false positives. Further, many of the potential serum tests fail to distinguish between benign and malignant disease. For example, CA125 is frequently positive in people with benign pelvic disease, and CA19.9, a putative pancreas screening assay, is frequently positive in people with pancreatitis. Imagine needing to tell a patient that you might have a cancer or a benign disease, but we do not know where it might be, whether it is a cancer or not, and what we should do about it. It is important to understand that there is a very high bar associated with the development of a successful screening assay. It is also critical to realize that even though a "test" might appear useful on a retrospectively and carefully collected sample set, prospective evaluations in a general environment will be necessary to determine utility. For example, a trial involving 150,000 women is currently ongoing to determine whether measuring CA125 over time combined with ultrasound will alter the outcomes for women at risk for ovarian cancer.

Based on the intriguing early evidence that particular lysophospholipids are elevated in the plasma and serum of specific cancer patients (even though these are controversial), and the critical importance of having effective screening approaches. We propose that it will be necessary to develop a national or international consortium to determine whether practical assays can be developed and whether these will be useful in patient management. It is important that the consortium adopt rigorous criteria for assay development and implementation, and that the process to move from assay development to clinical utility be carefully planned. The EDRN (Early Diagnosis Research Network) (Mills *et al.*, 2001) has developed a clear set of steps and criteria for the development and validation of lysophospholipids as potential markers for cancer development. We await the development, validation, and assessment of such assays.

ACKNOWLEDGMENTS

Supported by the U.S. Department of Defense (grants DAMD17-02-1-0691 1 and DAMD17-03-1-0409 01), and the National Institutes of Health [NIH] (Ovarian Cancer PPG grant P01 CA64602, and AVON-NCI Ovarian SPORE grant P50 CA083639). M.M. was supported by a training fellowship from the Keck Center Pharmacoinformatics Training Program of the Gulf Coast Consortia (NIH grant 1 T90 070109-01). We would also like to acknowledge the support of the Biosample Repository Core Facility at Fox Chase Cancer Center.

REFERENCES

Alpturk, O., Rusin, O., Fakayode, S. O., Wang, W., Escobedo, J. O., Warner, I. M., Crowe, W. E., Kral, V., Pruet, J. M., and Strongin, R. M. (2006). Lanthanide complexes as fluorescent indicators for neutral sugars and cancer biomarkers. *Proc. Natl. Acad. Sci. USA* **103,** 9756–9760.

American Cancer Society (2006). "Facts and Figures." American Cancer Society, Atlanta, GA.

Argraves, K. M., Wilkerson, B. A., Argraves, W. S., Fleming, P. A., Obeid, L. M., and Drake, C. J. (2004). Sphingosine-1-phosphate signaling promotes critical migratory events in vasculogenesis. *J. Biol. Chem.* **279,** 50580–50590.

Baker, D. L., Desiderio, D. M., Miller, D. D., Tolley, B., and Tigyi, G. J. (2001). Direct quantitative analysis of lysophosphatidic acid molecular species by stable isotope dilution electrospray ionization liquid chromatography-mass spectrometry. *Anal. Biochem.* **292,** 287–295.

Baker, D. L., Morrison, P., Miller, B., Riely, C. A., Tolley, B., Westermann, A. M., Bonfrer, J. M., Bais, E., Moolenaar, W. H., and Tigyi, G. (2002). Plasma lysophosphatidic acid concentration and ovarian cancer. *JAMA* **287**, 3081–3082.

Baker, D. L., Fujiwara, Y., Pigg, K. R., Tsukahara, R., Kobayashi, S., Murofushi, H., Uchiyama, A., Murakami-Murofishi, K., Koh, E., Bandle, R. W., Byun, H. S., Bittman, R., *et al.* (2007). Carba analogs of cyclic phosphatidic acid selective inhibitors of autotaxin and cancer cell invasion and metastasis. *J.Biol. Chem.* **281**, 22786–22793.

Bartlett, G. R. (1958). Phosphorous assay in a column chromatography. *J. Biol. Chem.* **234**, 466–468.

Boucharaba, A., Serre, C. M., Guglielmi, J., Bordet, J. C., Clezardin, P., and Peyruchaud, O. (2006). The type 1 lysophosphatidic acid receptor is a target for therapy in bone metastases. *Proc. Natl. Acad. Sci. USA* **103**, 9643–9648.

Ishida, M., Imagawa, M., Shimizu, T., and Taguchi, R. (2005). Specific detection of lysophosphatidic acids in serum extracts by tandem mass spectrometry. *J. Mass Spectrom. Soc. Jpn.* **53**, 35–51.

Johnson, K. R., Johnson, K. Y., Crellin, H. G., Ogretmen, B., Boylan, A. M., Harley, R. A., and Obeid, L. M. (2005). Immunohistochemical distribution of sphingosine kinase 1 in normal and tumor lung tissue. *J. Histochem. Cytochem.* **53**, 1159–1166.

Kawamori, T., Osta, W., Johnson, K. R., Pettus, B. J., Bielawski, J., Tanaka, T., Wargovich, M. J., Reddy, B. S., Hannun, Y. A., Obeid, L. M., and Zhou, D. (2006). Sphingosine kinase 1 is up-regulated in colon carcinogenesis. *FASEB J.* **20**, 386–388.

Kehlen, A., Englert, N., Seifert, A., Klonisch, T., Dralle, H., Langner, J., and Hoang-Vu, C. (2004). Expression, regulation and function of autotaxin in thyroid carcinomas. *Int. J. Cancer* **109**, 833–838.

Kishi, Y., Okudaira, S., Tanaka, M., Hama, K., Shida, D., Kitayama, J., Yamori, T., Aoki, J., Fujimaki, T., and Arai, H. (2006). Autotaxin is overexpressed in glioblastoma multiforme and contributes to cell motility of glioblastoma by converting lysophosphatidylcholine to lysophosphatidic acid. *J. Biol. Chem.* **281**, 17492–17500.

Kishimoto, T., Matsuoka, T., Imamura, S., and Mizuno, K. (2003). A novel colorimetric assay for the determination of lysophosphatidic acid in plasma using an enzymatic cycling method. *Clin. Chim. Acta* **333**, 59–67.

Kishimoto, T., Soda, Y., Matsuyama, Y., and Mizuno, K. (2002). An enzymatic assay for lysophosphatidylcholine concentration in human serum and plasma. *Clin. Biochem.* **35**, 411–416.

Kitayama, J., Shida, D., Sako, A., Ishikawa, M., Hama, K., Aoki, J., Arai, H., and Nagawa, H. (2004). Over-expression of lysophosphatidic acid receptor-2 in human invasive ductal carcinoma. *Breast Cancer Res.* **6**, R640–R646.

Kohno, M., Momoi, M., Oo, M. L., Paik, J. H., Lee, Y. M., Venkataraman, K., Ai, Y., Ristimaki, A. P., Fyrst, H., Sano, H., Rosenberg, D., Saba, J. D., *et al.* (2006). Intracellular role for sphingosine kinase 1 in intestinal adenoma cell proliferation. *Mol. Cell Biol.* **26**, 7211–7223.

Le Scolan, E., Pchejetski, D., Banno, Y., Denis, N., Mayeux, P., Vainchenker, W., Levade, T., and Moreau-Gachelin, F. (2005). Overexpression of sphingosine kinase 1 is an oncogenic event in erythroleukemic progression. *Blood* **106**, 1808–1816.

Lee, O. H., Kim, Y. M., Lee, Y. M., Moon, E. J., Lee, D. J., Kim, J. H., Kim, K. W., and Kwon, Y. G. (1999). Sphingosine 1-phosphate induces angiogenesis: Its angiogenic action and signaling mechanism in human umbilical vein endothelial cells. *Biochem. Biophys. Res. Commun.* **264**, 743–750.

Mills, G. B., Bast, R. C., Jr., and Srivastava, S. (2001). Future for ovarian cancer screening: Novel markers from emerging technologies of transcriptional profiling and proteomics. *J. Natl. Cancer Inst.* **93,** 1437–1439.

Milstien, S., and Spiegel, S. (2006). Targeting sphingosine-1-phosphate: A novel avenue for cancer therapeutics. *Cancer Cell* **9,** 148–150.

Nam, S. W., Clair, T., Kim, Y. S., McMarlin, A., Schiffmann, E., Liotta, L. A., and Stracke, M. L. (2001). Autotaxin (NPP-2), a metastasis-enhancing motogen, is an angiogenic factor. *Cancer Res.* **61,** 6938–6944.

Nava, V. E., Hobson, J. P., Murthy, S., Milstien, S., and Spiegel, S. (2002). Sphingosine kinase type 1 promotes estrogen-dependent tumorigenesis of breast cancer MCF-7 cells. *Exp. Cell Res.* **281,** 115–127.

Okita, M., Gaudette, D. C., Mills, G. B., and Holub, B. J. (1997). Elevated levels and altered fatty acid composition of plasma lysophosphatidylcholine(lysoPC) in ovarian cancer patients. *Int. J. Cancer* **71,** 31–34.

Sarkar, S., Maceyka, M., Hait, N. C., Paugh, S. W., Sankala, H., Milstien, S., and Spiegel, S. (2005). Sphingosine kinase 1 is required for migration, proliferation and survival of MCF-7 human breast cancer cells. *FEBS Lett.* **579,** 5313–5317.

Satouchi, K., Sakaguchi, M., Shirakawa, M., Hirano, K., and Tanaka, T. (1994). Lysophosphatidylcholine from white muscle of bonito *Euthynnus pelamis* (Linnaeus): Involvement of phospholipase A1 activity for its production. *Biochim. Biophys. Acta* **1214,** 303–308.

Saulnier-Blache, J. S., Girard, A., Simon, M. F., Lafontan, M., and Valet, P. (2000). A simple and highly sensitive radioenzymatic assay for lysophosphatidic acid quantification. *J. Lipid Res.* **41,** 1947–1951.

Shen, Z., Wu, M., Elson, P., Kennedy, A. W., Belinson, J., Casey, G., and Xu, Y. (2001). Fatty acid composition of lysophosphatidic acid and lysophosphatidylinositol in plasma from patients with ovarian cancer and other gynecological diseases. *Gynecol. Oncol.* **83,** 25–30.

Sliva, D., Mason, R., Xiao, H., and English, D. (2000). Enhancement of the migration of metastatic human breast cancer cells by phosphatidic acid. *Biochem. Biophys. Res. Commun.* **268,** 471–479.

Spiegel, S., and Milstien, S. (2003). Sphingosine-1-phosphate: An enigmatic signalling lipid. *Nat. Rev. Mol. Cell Biol.* **4,** 397–407.

Stracke, M. L., Krutzsch, H. C., Unsworth, E. J., Arestad, A., Cioce, V., Schiffmann, E., and Liotta, L. A. (1992). Identification, purification, and partial sequence analysis of autotaxin, a novel motility-stimulating protein. *J. Biol. Chem.* **267,** 2524–2529.

Sugo, T., Tachimoto, H., Chikatsu, T., Murakami, Y., Kikukawa, Y., Sato, S., Kikuchi, K., Nagi, T., Harada, M., Ogi, K., Ebisawa, M., and Mori, M. (2006). Identification of a lysophosphatidylserine receptor on mast cells. *Biochem. Biophys. Res. Commun.* **341,** 1078–1087.

Sutphen, R., Xu, Y., Wilbanks, G. D., Fiorica, J., Grendys, E. C., Jr., LaPolla, J. P., Arango, H., Hoffman, M. S., Martino, M., Wakeley, K., Griffin, D., Blanco, R. W., et al. (2004). Lysophospholipids are potential biomarkers of ovarian cancer. *Cancer Epidemiol. Biomarkers Prev.* **13,** 1185–1191.

Tanaka, T., Tsutsui, H., Hirano, K., Koike, T., Tokumura, A., and Satouchi, K. (2004). Quantitative analysis of lysophosphatidic acid by time-of-flight mass spectrometry using a phosphate-capture molecule. *J. Lipid Res.* **45,** 2145–2150.

Tanyi, J. L., Hasegawa, Y., Lapushin, R., Morris, A. J., Wolf, J. K., Berchuck, A., Lu, K., Smith, D. I., Kalli, K., Hartmann, L. C., McCune, K., Fishman, D., et al. (2003a). Role of decreased levels of lipid phosphate phosphatase-1 in accumulation of lysophosphatidic acid in ovarian cancer. *Clin. Cancer Res.* **9,** 3534–3545.

Tanyi, J. L., Morris, A. J., Wolf, J. K., Fang, X., Hasegawa, Y., Lapushin, R., Auersperg, N., Sigal, Y. J., Newman, R. A., Felix, E. A., Atkinson, E. N., and Mills, G. B. (2003b). The human lipid phosphate phosphatase-3 decreases the growth, survival, and tumorigenesis of ovarian cancer cells: Validation of the lysophosphatidic acid signaling cascade as a target for therapy in ovarian cancer. *Cancer Res.* **63,** 1073–1082.

Tokumura, A., Majima, E., Kariya, Y., Tominaga, K., Kogure, K., Yasuda, K., and Fukuzawa, K. (2002a). Identification of human plasma lysophospholipase D, a lysophosphatidic acid-producing enzyme, as autotaxin, a multifunctional phosphodiesterase. *J. Biol. Chem.* **277,** 39436–39442.

Tokumura, A., Tominaga, K., Yasuda, K., Kanzaki, H., Kogure, K., and Fukuzawa, K. (2002b). Lack of significant differences in the corrected activity of lysophospholipase D, producer of phospholipid mediator lysophosphatidic acid, in incubated serum from women with and without ovarian tumors. *Cancer* **94,** 141–151.

Umezu-Goto, M., Kishi, Y., Taira, A., Hama, K., Dohmae, N., Takio, K., Yamori, T., Mills, G. B., Inoue, K., Aoki, J., and Arai, H. (2002). Autotaxin has lysophospholipase D activity leading to tumor cell growth and motility by lysophosphatidic acid production. *J. Cell Biol.* **158,** 227–233.

Umezu-Goto, M., Tanyi, J., Lahad, J., Liu, S., Yu, S., Lapushin, R., Hasegawa, Y., Lu, Y., Trost, R., Bevers, T., Jonasch, E., Aldpe, K., *et al.* (2004). Lysophosphatidic acid productio anaction: Validated targets in cancer? *J. Cell Biochem.* **92,** 1115–1140.

Van Brocklyn, J. R., Jackson, C. A., Pearl, D. K., Kotur, M. S., Snyder, P. J., and Prior, T. W. (2005). Sphingosine kinase-1 expression correlates with poor survival of patients with glioblastoma multiforme: Roles of sphingosine kinase isoforms in growth of glioblastoma cell lines. *J. Neuropathol. Exp. Neurol.* **64,** 695–705.

van Meeteren, L. A., Ruurs, P., Christodoulou, E., Goding, J. W., Takakusa, H., Kikuchi, K., Perrakis, A., Nagano, T., and Moolenaar, W. H. (2005). Inhibition of autotaxin by lysophosphatidic acid and sphingosine 1-phosphate. *J. Biol. Chem.* **280,** 21155–21161.

van Meeteren, L. A., Ruurs, P., Stortelers, C., Bouwman, P., van Rooijen, M. A., Pradere, J. P., Pettit, T. R., Wakelam, M. J., Saulnier-Blache, J. S., Mummery, C. L., Moolenaar, W. H., and Jonkers, J. (2006). Autotaxin, a secreted lysophospholipase D, is essential for blood vessel formation during development. *Mol. Cell Biol.* **26,** 5015–5022.

Visentin, B., Vekich, J. A., Sibbald, B. J., Cavalli, A. L., Moreno, K. M., Matteo, R. G., Garland, W. A., Lu, Y., Yu, S., Hall, H. S., Kundra, V., Mills, G. B., *et al.* (2006). *Cacer Cell* **9,** 225–238.

Xiao, Y., Chen, Y., Kennedy, A. W., Belinson, J., and Xu, Y. (2000). Evaluation of plasma lysophospholipids for diagnostic significance using electrospray ionization mass spectrometry (ESI-MS) analyses. *Ann. NY Acad. Sci.* **905,** 242–259.

Xiao, Y. J., Schwartz, B., Washington, M., Kennedy, A., Webster, K., Belinson, J., and Xu, Y. (2001). Electrospray ionization mass spectrometry analysis of lysophospholipids in human ascitic fluids: Comparison of the lysophospholipid contents in malignant vs nonmalignant ascitic fluids. *Anal. Biochem.* **290,** 302–313.

Xu, Y., Fang, X. J., Casey, G., and Mills, G. B. (1995). Lysophospholipids activate ovarian and breast cancer cells. *Biochem. J.* **309,** 933–1040.

Xu, Y., Shen, Z., Wiper, D. W., Wu, M., Morton, R. E., Elson, P., Kennedy, A. W., Belinson, J., Markman, M., and Casey, G. (1998). Lysophosphatidic acid as a potential biomarker for ovarian and other gynecologic cancers. *JAMA* **280,** 719–723.

Yang, S. Y., Lee, J., Park, C. G., Kim, S., Hong, S., Chung, H. C., Min, S. K., Han, J. W., Lee, H. W., and Lee, H. Y. (2002). Expression of autotaxin (NPP-2) is closely linked to invasiveness of breast cancer cells. *Clin. Exp. Metastasis* **19,** 603–608.

Yang, Y., Mou, L., Liu, N., and Tsao, M. S. (1999). Autotaxin expression in non–small-cell lung cancer. *Am. J. Respir. Cell Mol. Biol.* **21,** 216–222.

Zhang, G., Zhao, Z., Xu, S., Ni, L., and Wang, X. (1999). Expression of autotaxin mRNA in human hepatocellular carcinoma. *Chin. Med. J. (Engl.)* **112,** 330–332.

CHAPTER TWO

MEASUREMENT OF EICOSANOIDS IN CANCER TISSUES

Dingzhi Wang* *and* Raymond N. DuBois[†]

Contents

Abstract

Eicosanoids derived from arachidonic acid through cyclooxygenase (COX), lipoxygenase (LOX), and P450 pathways include prostanoids, hydroxyeicosatetraenoic acids (HETEs), leukotrienes (LTs), and epoxyeicosatrienoic acids (EETs).

* Department of Medicine, Vanderbilt University Medical Center, Nashville, Tennessee
[†] Departments of Medicine, Cancer Biology, Cell and Developmental Biology, Vanderbilt University Medical Center, and Vanderbilt-Ingram Cancer Center, Nashville, Tennessee

Methods in Enzymology, Volume 433
ISSN 0076-6879, DOI: 10.1016/S0076-6879(07)33002-4

These bioactive lipids play an important role in regulating cell proliferation, apoptosis, tissue repair, blood clotting, blood vessel permeability, inflammation, and immune cell behavior. Moreover, some of these eicosanoids also modulate inflammation and tumor growth in cancer tissues, and may serve as biomarkers for monitoring colorectal cancer progression or as an intermediate marker for the pharmacologic activity of chemopreventive agents. Development of sensitive, rapid, and specific methods for determining eicosanoid levels accurately will facilitate an understanding of the biologic functions of these lipid mediators and will broaden our insight of the importance of these bioactive lipids *in vivo*. However, quantitative determination of eicosanoids in biological samples has presented a problem to many investigators. It is necessary to understand the advantages and limitations of each method for quantitative analysis of specific eicosanoids in various types of biological samples. Here we evaluate the methodology of the measurement of eicosanoids in biological samples.

1. INTRODUCTION

Eicosanoids derived from the arachidonic acid are bioactive lipids that play an important role in the pathogenesis of a variety of human diseases, including inflammation and cancer. Arachidonic acid (AA) is a polyunsaturated fatty acid that constitutes the phospholipid domain of most cell membranes. When tissues are exposed to diverse physiologic and pathologic stimuli, arachidonic acid is liberated from membrane phospholipids by action of cytoplasmic phospholipase A_2. Arachidonic acid can be metabolized to eicosanoids through three major pathways: the cyclooxygenase (COX) pathway, the lipoxygenase (LOX) pathway, and the cytochrome P-450 monooxygenase pathway. Clinical and epidemiologic studies have already demonstrated the importance of COX and LOX in the progression of colorectal cancer. Moreover, both pharmacologic and genetic evidence show that some of eicosanoids regulate colorectal tumor growth and are involved in other types of cancer as well (Hussey and Tisdale, 1994; Hussey *et al.*, 1996; Wang *et al.*, 2005). Therefore, these bioactive lipids may serve as a biomarker for monitoring cancer progression and a valuable intermediate marker for the pharmacological activity of chemopreventive agents. Successful quantitation of eicosanoids in biological samples is challenging to many investigators. It is vital to understand the advantages and limitations of each method for quantitative analysis of specific eicosanoids in different biological samples such as intestinal tissue, plasma/serum, urine, and cell culture.

1.1. COX pathway

Metabolism of free arachidonic acid (AA) by cyclooxygenase enzymes leads to the formation of prostanoids, including prostaglandins (PGs) and thromboxanes (TXs) (DuBois *et al.*, 1998; Herschman, 1996; Herschman *et al.*, 1995; Smith *et al.*, 2000). The key regulatory step in this process is the enzymatic conversion of the AA to PGG_2, which is then reduced to an unstable endoperoxide intermediate, PGH_2. PGH_2 is sequentially metabolized to five active, structurally related prostanoids, including PGE_2, PGD_2, $PGF_{2\alpha}$, PGI_2, and thromboxane A_2 (TxA_2), in a cell type–specific manner via specific PG synthases (Fig. 2.1). These bioactive lipids exert their cellular functions by binding cell surface receptors that belong to the family of seven transmembrane G-protein–coupled rhodopsin-type receptors. These receptors are designated DP for the PGD_2 receptor, EP (EP1, EP2, EP3, and EP4) for PGE_2 receptors, FP for the $PGF_{2\alpha}$ receptor, IP for the PGI_2 receptor, and TP for the TXA_2 receptor. Moreover, some PGs can also bind to nuclear receptors such as peroxisome proliferators–activated receptors (PPARs). It has been shown that the PGD_2 dehydration product 15-deoxy-$^{\Delta12, \Delta14}PGJ_2$ (15dPGJ$_2$) is a natural ligand for the PPARγ receptor (Forman *et al.*, 1995; Kliewer *et al.*, 1995), while PGI_2 activates PPARδ by directly binding to this receptor (Forman *et al.*, 1997; Gupta *et al.*, 2000). Additionally, PGE_2 has been shown to indirectly transactivate PPARδ

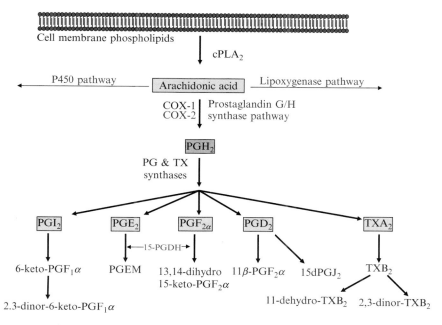

Figure 2.1 Overview of PGHS (COX) pathway. (See color insert.)

(Wang *et al.*, 2004). Recent studies suggest that PPARγ and PPARδ may play an important role in modulating colorectal carcinogenesis as well as other types of cancer (Cellai *et al.*, 2006; Gupta *et al.*, 2004; Hussey and Tisdale, 1994; Hussey *et al.*, 1996; Panigrahy *et al.*, 2005; Takayama *et al.*, 2006; Wang *et al.*, 2004; Yin *et al.*, 2005).

COX enzymes (more correctly referred to as prostaglandin G/H synthases) exist in two isoforms: COX-1 (PGHS-1) and COX-2 (PGHS-2). In the gut, most studies indicate that COX-1–derived PGs are produced by epithelial and stromal cells in subepithelial tissue (Craven and DeRubertis, 1983; Lawson and Powell, 1987; Smith *et al.*, 1982). COX-1–derived PGs play an important role in protecting the gastroduodenal mucosa from damage. For example, PGI$_2$ is a major product of prostaglandins (LeDuc and Needleman, 1980; Lee *et al.*, 1992) and plays a key role in the cytoprotection of gastric mucosal surfaces and the normal vasculature (McAdam *et al.*, 1999). In addition, PGI$_2$ appears to serve an important role in protecting cardiomyocytes from oxidant stress (Adderley and Fitzgerald, 1999). In contrast, COX-2 derived PGs not only mediate acute inflammatory responses such as swelling, pain, and fever, but are also involved in a variety of pathophysiological processes, including colorectal cancer, ulceration, thrombosis, and kidney disease. For example, PGE$_2$ is the most abundant PG found in human colorectal cancer (Rigas *et al.*, 1993). A recent study showed that PGE$_2$ treatment dramatically increased both small and large intestinal adenoma burden in $Apc^{Min/+}$ mice and significantly enhanced colon carcinogen (AOM)–induced colon tumor incidence and multiplicity (Kawamori *et al.*, 2003; Wang *et al.*, 2004). Furthermore, PGE$_2$ protects small intestinal adenomas from NSAID-induced regression in $Apc^{Min/+}$ mice (Hansen-Petrik *et al.*, 2002). The central role of PGE$_2$ in colorectal tumorigenesis has been further confirmed by evaluating mice with homozygous deletion of PGE$_2$ receptors (Mutoh *et al.*, 2002; Sonoshita *et al.*, 2001; Watanabe *et al.*, 1999). In addition, PGE$_2$ is also a key mediator of acute inflammatory responses (Needleman and Isakson, 1997; Portanova *et al.*, 1996), arthritis (Amin *et al.*, 1999; Anderson *et al.*, 1996), and inflammatory bowel disease (Gould *et al.*, 1981; MacDermott, 1994).

1.2. LOX pathway

LOXs convert arachidonic, linoleic, and other polyunsaturated fatty acids into bioactive metabolites such as leukotrienes (LTs), hydroxyeicosatetraenoic acids (HETEs), and hydroxyoctadecadienoic acid (HODE). Although most attention has focused on COX-derived PGs, emerging evidence suggests that LOX-catalyzed products, LTs, HETEs, and HODE, also exert profound biological effects on the development and progression of human cancers, including colorectal cancer. In general, 5-, 8-, and 12-LOX has potential procarcinogenic roles while 15-LOX-1 has anticarcinogenic

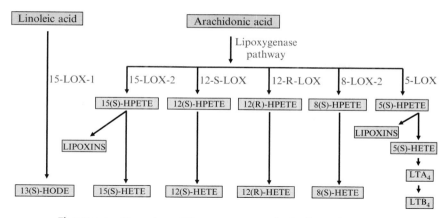

Figure 2.2   Overview of lipoxygenase synthesis. (See color insert.)

effect (Shureiqi and Lippman, 2001; Shureiqi *et al.*, 2005). In an eicosanoid-generation pathway, arachidonic acid was found to serve as a substrate for 5-, 8-, 12-, and 15-LOX which catalyze the stereospecific oxygenation (oxidation) of the 5-, 8-, 12-, or 15-carbon atoms of arachidonic acid to generate the corresponding hydroperoxyeicosatetraenoic acids (HPETEs), respectively. HPETEs are sequentially metabolized into the corresponding HETEs, LTs, or lipoxins in a cell type–specific manner (Fig. 2.2). Several LOXs and their metabolites of arachidonic acid appear to be involved in colon carcinogenesis. For example, LTB_4, 5(S)-HETE, 12(S)-HETE, and 12(R)-HETE stimulate colon cancer tumor growth *in vitro* and *in vivo* (Bortuzzo *et al.*, 1996; Hussey and Tisdale, 1994; Hussey *et al.*, 1996; Ye *et al.*, 2005). In contrast, 15-HETE and 13-HODE appear to inhibit the growth of colon and prostate carcinoma cells (Bhatia *et al.*, 2003; Shureiqi *et al.*, 1999).

1.3. Cytochrome P-450 pathway

Arachidonic acid (AA) is also metabolized by cytochrome P-450 to AA epoxygenase products (epoxyeicosatrienoic acids), AA ω/ω-1 hydroxylase products (hydroxyeicosatetraenoic acids), lipoxygenase-like products (hydroxyeicosatetraenoic acids), and free radical oxidation products (hydroperoxyeicosatetraenoic acids) (Fig. 2.3). In mammals, 14 families and 26 subfamilies of cytochromes P-450 (CYPs) have been identified. The expression of cytochrome P450s has been identified in a wide range of human cancers (Patterson and Murray, 2002). For example, CYP1B1 is elevated in tumors including lung, breast, liver, gastrointestinal tract, prostate, and bladder cancers. The major cytochrome P450-generated metabolites are the epoxyeicosatrienoic

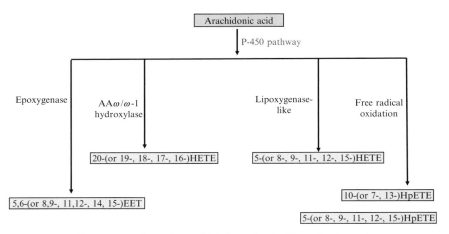

Figure 2.3   Overview of P-450 synthesis. (See color insert.)

acids (EETs) (including 5,6-EET, 8,9-EET, 11,12-EET, and 14,15-EET) and the hydroxyeicosatetraenoic acids (HETEs) (mainly 20–HETE and 19-HETE) in humans (Spector *et al.*, 2004). These products play an important role in the regulation of inflammation, cell migration, apoptosis, and platelet aggregation (Capdevila *et al.*, 2000; Potente *et al.*, 2003). For example, EETs activate EGFR, MAP kinase, and PI3 kinase signaling pathways in endothelial and epithelial cells (Chen *et al.*, 1998, 1999; Fleming *et al.*, 2001; Hoebel and Graier, 1998).

2. QUANTITATIVE MEASUREMENT OF EICOSANOIDS

It is critical to evaluate the methodology used for the measurement of eicosanoids, particularly for PGE_2, because its concentration in the supernatant of colorectal cancer cell lines and in intestinal tissues, plasma, and urine from both humans and mice may serve as a biomarker for monitoring colorectal cancer progression and a valuable intermediate marker for the pharmacological activity of chemopreventive agents. The direct assay of PGE_2 from the culture medium is a good way to measure PGE_2 production from cultured cells. However, PGs including PGE_2 are unstable compounds that are rapidly metabolized *in vivo* (see Fig. 2.1). Therefore, the direct quantitation of PGs is an unreliable indicator in an *in vivo* system. For example, PGE_2 and PGF_2 are rapidly metabolized to a stable 13,14-dihydro-15-keto-PGA_2 (PGEM) and 13,14-dihydro-15-keto-PGF_2 *in vivo* by the enzyme 15-hydroxy prostaglandin dehydrogenase (15-PGDH), respectively (Ensor *et al.*, 1990; Keirse and Turnbull, 1975; Samuelsson *et al.*, 1975; see Fig. 2.1). In addition, PGI_2 is nonenzymatically hydrated to 6-keto PGF_1, and then quickly converted to the major urinary metabolite, 2,3-dinor-6-keto PGF_1 (Rosenkranz *et al.*, 1980; Samuelsson *et al.*, 1978), and PGD_2 is also rapidly metabolized to 15d-PGJ_2 and 11β-PGF_2 *in vivo* (Liston and Roberts, 1985). Moreover, TXA_2 is rapidly hydrolyzed to TXB_2 in a nonenzymatic manner. TXB_2 can be further metabolized by 11-hydroxy thromboxane dehydrogenase to 11-dehydro TXB_2, or by β-oxidation to form 2,3-dinor TXB_2 (Roberts *et al.*, 1981). For this reason, blood, urine, colorectal tissue, or other samples from humans and animals often contain very few intact PGs, and measurement of PG metabolites is crucial to provide a reliable estimate of actual PGs production in the samples that have undergone extensive metabolism before collection (Catella *et al.*, 1986; Del Vecchio *et al.*, 1992; Frolich, 1984; Meyer *et al.*, 1989; O'Sullivan *et al.*, 1996, 1999). In addition, it is essential to determine the percentage recovery of each eicosanoid in biological samples when any extraction and/ or purification procedures are employed.

Eicosanoids can be quantitatively analyzed by spectrophotometry, radio-immunoassay (RIA), enzyme immunoassay (EIA), gas chromatography–mass spectrometry (GC-MS), and liquid chromatography–mass spectrometry–mass spectrometry (LC-MS-MS). The spectrophotometric method is relatively specific and simple, but has serious sensitivity limitations, which is limited to the low microgram range (Bygdeman and Samuelsson, 1964; Shaw and Ramwell, 1969). The RIA and EIA are sensitive enough to measure subpicomole amounts of eicosanoids, but tedious. However, these assays have two limitations for tissue and plasma samples. First, plasma proteins can bind to eicosanoids, which diminishes immunoassay sensitivity. Second, there is a significant degree of immunological cross-reactivity among commercially available eicosanoid antibodies. To solve these problems, an organic solvent or a solid-phase extraction procedure is used to separate eicosanoids from plasma proteins, and chromatographic separation of eicosanoids is employed to avoid immunological cross-reactivity (Jaffe *et al.*, 1973). The GC-MS and LC-MS-MS are specific and relatively sensitive at the picogram range. In addition, these methods are very useful for the measurement of products in the eicosanoid degradation pathway (Green *et al.*, 1973; Thompson *et al.*, 1970).

3. PREPARATION OF SAMPLES

The amount of sample depends on the nature of the sample matrix and the relatively minimal levels of eicosanoids produced by isolated cell suspensions to more complex matrixes such as intestinal tissues, urine, and blood. Sample preparation also depends on the method used for quantitative analysis of eicosanoids. Polypropylene tubes are used throughout the following process to avoid binding of the eicosanoids to glass surfaces.

3.1. Urine samples

In general, 1 to 3 ml of urine is collected for a single measurement. Indomethacin should be added immediately to urine samples at a 10 μM final concentration. Indomethacin will prevent *ex vivo* formation of eicosanoids. The samples are centrifuged in a microcentrifuge (12,000 rpm for 2 min at 4°) to remove any precipitate. A "hot" spike of eicosanoids of interest is added to determine the percent recovery of the samples during extraction and/or purification procedures as described later. The samples are acidified to pH 3 by adding glacial acetic acid. Organic acids are suggested for this purpose since mineral acids promote the dehydration of PGE_2 to PGA_2 at a faster rate.

3.2. Plasma preparation

In general, 1 to 2 ml of blood is collected in polypropylene tube coated with a spray-dried anticoagulant such as sodium heparin using plastic syringes. Indomethacin should be added immediately to blood samples as described previously. The plasma samples are obtained by centrifuging in a microcentrifuge (4000 rpm for 30 min at 4°). A "hot" or "cold" spike of eicosanoids of interest is added as described later. The samples are acidified to pH 3 by adding glacial acetic acid.

3.3. Collection of intestinal tumor samples

Tumor and matched normal tissue samples (range, 100 to 500 mg) are collected from human colon or mouse intestines and immediately snap frozen in liquid nitrogen with approximately the same time period between biopsy removal and freezing for each biopsy. All tissue samples should be stored in liquid nitrogen until the time of extraction.

3.4. Tissue homogenization

Frozen tissue (50 to 500 mg) is ground to a fine powder using a liquid-nitrogen–cooled mortar (Fisher) and homogenized in five volumes of ice-cold PBS buffer with 0.1% butylated hydroxytoluene and 1 mM EDTA and 10 μM indomethacin. Tissue is homogenized by hand in a siliconized Duall glass–glass tissue grinder (Kontes Glass, Vineland, CT) for approximately 30 s or until there are no visible pieces of tissue or by an Ultrasonic Processor (Misonix, Farmingdale, NJ) at 0° for 3 min. The homogenate is added to three volumes of 100% ethanol, vortexed, and put on ice for

5 min. The samples are centrifuged in a microcentrifuge (12,000 rpm for 2 min at 4°) to remove any protein precipitate. Then double-deionized water is added to tissue homogenate to reach a final concentration of 15% ethanol. Similarly, a "hot" spike of eicosanoids is added to the samples. The mixtures are centrifuged at 400g for 10 min, and the supernatants were transferred to a fresh tube. The samples are acidified to pH 3 by adding glacial acetic acid.

4. Eicosanoids Extraction

Eicosanoids can be extracted from biological samples using organic solvent or octadecylsilyl (ODS) silica. The most common method for extracting eicosanoids and eicosanoid metabolites from acidified aqueous solutions is an organic solvent extraction. However, extraction on a solid-phase, such as ODS silica, column is a simple chromatographic method with relative selection and small volumes of the extraction (Powell, 1980).

4.1. Organic solvent extraction

Ten thousand counts per minute (cpm) of tritium-labeled eicosanoids of interest are added to the samples (1 to 5 ml of urine or cell-free tissue culture medium, 1 to 3 ml of plasma, or 1 to 5 ml of the supernatant of tissue homogenate), acidified to pH 3 with glacial acetic acid and the lipids are extracted twice with five volumes of diethyl ether and once with five volumes of ethyl acetate. Recovery values are about 95% after extraction based on the scintillation counting. The organic phase is evaporated to dryness under a stream of nitrogen at room temperature. The dry extract is stored at −80° until the time of the experiment. The dry extract is reconstituted in an appropriate assay buffer for the following procedures.

4.2. Extraction using an octadecylsilyl silica column

A 6-ml SEP-PAK C18 cartridges (Waters Associates) or 6-ml ODS silica column (J. T. Baker, #7020-6) is activated by rinsing with 100% ethanol (20 ml) followed by double-deionized water (20 ml) to remove excess ethanol. The acidified samples with 10,000 cpm of tritium-labeled eicosanoids of interest are passed through the prepared cartridge or column using a syringe or under gentle vacuum with a Vac-Elut (Vanian Sample Preparation Products, Harbor City, CA). For larger samples (incubation volumes up to 25 ml), it may be advisable to use more than one ODS silica cartridge or an open column of ODS silica. Body fluids such as plasma and urine can be applied directly to the prepared cartridge or column after acidification. The cartridge or column is then washed with distilled water (20 ml) followed by 15% aqueous ethanol. If the sample applied to the ODS silica is dissolved in 15% aqueous ethanol, many polar materials (e.g., phospholipids) will pass through the column without being retained. For the samples in aqueous media from the body fluids or cell culture medium, more polar materials will be removed by the subsequent elution with 15% aqueous ethanol. Eicosanoids can be eluted from ODS silica with organic solvents such as diethyl ether, ethyl acetate, and methyl formate in a siliconized glass or polypropylene tube. Since methyl formate is more polar than diethyl ether and ethyl acetate, only 10 ml are required to elute eicosanoids from the stationary phase. Larger volumes (e.g., 20 ml of diethyl ether) of less polar solvents are required for good recoveries. In addition, methyl formate (boiling point 32°) is a very volatile solvent, enabling it to be removed rapidly under a stream of nitrogen. After extraction, the recoveries of 6-keto-PGF$_{1\alpha}$, TxB$_2$, PGE$_2$, PGF$_{2\alpha}$, LTB$_4$, LTC$_4$, 5-HETE, 12-HETE, and 15-HETE are 90.5, 90.6, 92.5, 98.1, 86.1, 98.3, 95.3, 99.8, and 92.8%, respectively, based on the scintillation count (Hollenberg *et al.*, 1994). The dry extract is stored at $-80°$ until the time of measurement. The dry extract is reconstituted in an appropriate assay buffer. The column can be reused up to 10 times when extracting culture media and two to three times when extracting urine without contamination or a decrease in extraction efficiency, respectively. For extracting plasma, the column can be used only once.

5. Chromatographic Separation of Eicosanoids

Thin-layer chromatography (TLC) is the most commonly used technique for the separation of arachidonic acid (AA) metabolites. However, this method has low and variable recovery and requires the use of several solvent systems to adequately separate the major AA metabolites. Silicic acid column chromatography is a simple chromatographic method but suffers from poor resolution. High-pressure liquid chromatography (HPLC) offers the advantage of high resolution and good reproducibility.

5.1. Thin-layer chromatography

Unmodified thin-layer silica–coated plates have been used to separate PGs (Amin, 1975; Eastman and Dowsett, 1976; Green and Samuelsson, 1964). Recently, thin-layer silica plates coated with phenylmethylvinylchlorosilane have been employed to separate PGs and HETEs (Beneytout *et al.*, 2005).

No adequate thin-layer chromatography system is presently available for the separation of LTC_4 and LTD_4.

The dried sample extract is dissolved in 100 μl of chloroform:methanol (95:5) and spotted on the unmodified thin-layer silica coated plate (Merck, 60F254, 20 × 20 mm) under a stream of N_2. One sample is spotted per lane. The standards (5 μg of each PG) are spotted in the outer lanes. A lane is skipped between the standards and the samples to avoid contamination. Samples are spotted next to each other on the inner lanes. The plate is developed with a solution of ethyl acetate-benzene-acetic acid (24:12:1) for 30 to 40 min. The plates are allowed to dry for 5 to 10 min and placed in a chamber containing crystals of iodine until the PG standards can be visualized. The zones of migration of the unknown samples are marked lightly on the silica gel with a pencil. Glass wool plugs are placed in the bottom of Pasteur pipettes. A pipette is placed tip first into a rubber hose that is connected to a vacuum. The zone corresponding to the PG of interest is scraped with a metal spatula. The silica gel is vacuumed into the pipette as it is scraped. After removing all of the loose silica gel, the pipette is removed from the vacuum line and placed tip down in a labeled 20-ml polypropylene tube. This process is repeated for each PG zone of interest. The PGs are eluted from the silica gel by adding two 2-ml aliquots of chloroform: methanol (95:5) to the Pasteur pipette. The eluate is collected and evaporated to dryness at 35° under N_2. The sample is reconstituted in an appropriate assay buffer depending on the method of quantitative analysis. The recovery of each PG is determined by the scintillation counting.

For separating a mixture of PGs and HETEs, each sample and 5 μl ($\mu g/\mu l$) of standard (PGE_2, PGD_2, $PGF_{2\alpha}$, and $PGF_{2\beta}$, 6-keto-$PGF_{1\alpha}$, thromboxane B2, and hydroxyeicosatetraenoic acids (9 HETE, 11 HETE, 12 HETE, and 15 HETE) is spotted per lane on the phenylmethylvinylchlorosilane coated plates as described previously. The plate is developed with a solution of heptane:methyl formate:diethyl ether:acetic acid (65:25:10:2) for HETEs. The distance traveled by solvent is 8.5 cm. PGs and thromboxane are separated after two migrations (heptane:methyl formate:diethyl ether:acetic acid, 65:25:102 and hexane:methyl formate:diethyl ether:acetic acid, 50:40:10:1). The distance traveled by first migration is 8.5 cm, while the distance traveled by second migration is 11 cm. The samples are recovered from the plate as described previously.

5.2. Silicic acid column chromatography

Silicic acid (Bio-Sil A, 200 to 400 mesh, Bio-Rad, #131-1350) is suspended to 0.25 g/ml in benzene:ethyl acetate:methanol (60:40:3) (solvent 1). A glass wool plug is inserted into the bottom of glass tulip columns (Radnoti, 8 × 150 mm, #120830). The silicic acid is pipetted into the column to a level of 100 mm and the column is sequentially rinsed with 6 ml of solvent 1,

6 ml of solvent 2 (toluene:ethyl acetate:methanol:water, 60:40:20:1), and 4 ml of solvent 1. The dry sample extract is dissolved in 0.3 ml of solvent 3 (toluene:ethyl acetate:methanol:glacial acetic acid, 60:40:5:0.01) and then the tube is rinsed with 0.8 ml of solvent 1. The combined organic extracts are then applied to the columns. Prostaglandin fractions are obtained by eluting the columns serially with 10 ml of solvent 1 (fraction I, PGA_2 and PGB_2), 12 ml of solvent 4 (toluene: ethyl acetate: methanol: water, 60: 40: 4.5: 0.1) (fraction II, PGE_2, PGD_2, TxB_2, and 6-keto-$PGF_{2\alpha}$), and 10 ml of solvent 2 (fraction III, PGF_2). The fractions are taken to dryness by evaporation under N_2 at 35° (Campbell and Ojeda, 1987; Jaffe et al., 1973).

5.3. High-pressure liquid chromatography

Various reverse phase columns such as the Ultrasphere ODS column (5 μm, 4.6 × 250 mm, Beckman Instruments) (Brouard and Pascaud, 1990; Campbell and Ojeda, 1987), 8 mm × 10 cm Nova-Pak C18 cartridge (Waters Association, Milford, MA, USA) (Sanduja et al., 1991), or a μ-Bondapak "fatty acid analysis" column (Waters) (Ohm et al., 1984) have been used to separate eicosanoids by using microprocessor–controlled gradient liquid chromatography (Beckman-Altex, model 334) or a Varian Vista 5500 liquid chromatograph system (Varian Associate Inc, Walnut Creek, CA) with a variable wavelength (190 to 700 mm) detector. In general, the column is equilibrated for at least 10 min under 26% solvent B (0.05% glacial acetic acid and 0.1% benzene in acetonitrile) in solvent A (0.05% glacial acetic acid in water) at a flow rate of 3 ml/min. The sample with hot tracer and appropriate standards are usually dissolved in 50 to 200 μl of acetone and are injected into the column. Tritiated eicosanoids of interest are run in parallel as standards and used as references for the determination of retention times. For example, the elution times of the eicosanoids are calculated from the [³H] eicosanoid standards. [³H] eicosanoid standard (2000 cpm) is injected onto the column. The column eluate is collected in 0.5-ml fractions, and the radioactivity is determined by liquid scintillation counting. PGs are eluted in the initial isocratic elution with 31% solvent B in solvent A at 1 ml/min of flow rate. The detection wavelength is 193 nm. The column eluate is collected in fractions corresponding to the elution times of the [³H] PGs of interest. The recovery values for PGs vary between 82 and 95% (Ohm et al., 1984). After the initial isocratic period, a slightly convex gradient is run from 26 to 50% solvent B in solvent A. The LOX arachidonic acid metabolite HETEs are eluted with 50% solvent B in solvent A and collected in fractions corresponding to the elution times of the [³H] HETEs of interest. The mobile phase for leukotriene analysis is water:methanol:acetonitrile:acetic acid (51:9:39:1, v/v) adjusted to pH 5.6 with triethylamine. The flow rate is 1.0 ml/min and the detection wavelength is 280 nm. For P-450 arachidonic acid metabolites, epoxygenase and

AA ω/ω-1 hydroxylase products are eluted with water:acetonitrile:acetic acid (37.5:62.5:0.05) increased to acetonitrile by 1.88%/min at a flow rate of 1 ml/min. The column fractions are frozen and lyophilized in a Speed-Vac centrifuge (Savant). Alternatively, the eicosanoids from the column fractions can also be extracted by organic solvent or solid-phase extraction procedure as described previously. The dried fractions are redissolved in an appropriate assay buffer based on the subsequent method of eicosanoid analysis employed (Eling *et al.*, 1982).

6. QUANTITATIVE ANALYSIS OF EICOSANOIDS

Quantitative determination of eicosanoids can be challenging for some investigators. RIA is rapid and possesses the sensitivity and specificity with lower costs. Similarly, EIA is sensitive and specific, but the cost is much higher. However, antibodies to all the known AA metabolites are not readily available, which limits these techniques. GC-MS and LC-MS-MS techniques are sensitive and reliable, but the cost is very high. Therefore, these types of analysis methods are restricted to a few research laboratories.

6.1. Radioimmunoassay

Radioimmunoassay (RIA) has been used to measure the levels of eicosanoids in biological samples. The assay is based on the competition of antigen (the eicosanoid being measured) with a constant amount of radioactive homologous antigen for a limited amount of specific eicosanoid antibody. Thus, the necessary reagents are a monospecific antibody, standard unlabeled eicosanoid, and radioactive eicosanoid. In addition, a method of separating antibody-antigen complex from free antigen is required. For more detailed information on the analysis of eicosanoids by RIA, the reader is directed to several excellent reviews (Campbell and Ojeda, 1987; Hollenberg *et al.*, 1994; Salmon, 1983).

In general, biological samples are extracted by mixing with an organic solvent or solid-phase extraction and purified by reverse-phase HPLC or TLC as described previously. A standard curve is included in each assay of eicosanoid in duplicate. Fifty microliters of each appropriately diluted antibody in RIA buffer (0.01 M phosphate, pH 7.54, 0.15 M NaCl, and 0.1% gelatin), 8,000 to 10,000 cpm of tritiated eicosanoid (New England Nuclear Corp.), and 1 mM EDTA are incubated for 1 h at 4° with either unlabeled eicosanoid (15 to 4000 pg) as standard or the unknown sample dissolved in 100 μl of RIA buffer. Before separation of antibody-bound from unbound [^{3}H] eicosanoid, 100 μl of phosphate-buffered saline (PBS) containing 5 mg/ml gelatin is added to prevent nonspecific binding and stabilize the separation procedure (if added earlier, gelatinization prevents adequate reaction). The simplest and cheapest method of separating bound from free antigen is the adsorption of the free eicosanoid on dextran-coated charcoal (DCC). First, 1.0 ml of dextran-coated charcoal (2.5 mg/ml of activated Norit A charcoal and 0.25 mg/ml of dextran T70 in PBS, Pharmacia Fine Chemicals Inc., Piscataway, NJ) is added to the reaction vials. After a 10-min centrifugation at 4°, 3000 rpm, the supernatants (containing antibody bound [^{3}H] eicosanoid) are immediately decanted into 5 ml of liquid scintillation fluid, and the radioactivity is determined by liquid scintillation spectrometry. Because the charcoal absorbs the unbound eicosanoid, the radioactivity measured represents the antibody-bound eicosanoid: percentage bound = [cpm (tube) − cpm (nonspecific binding {NSB})/cpm (total)] × 100. Each eicosanoid standard curve is calculated by plotting the percentage of [^{3}H] eicosanoid bound to the antibody versus picograms of added eicosanoid standard. The concentration of eicosanoid in biological samples is determined by plotting the percentage bound of unknown sample against the standard curve. Recently, RIA kits for several eicosanoids have become commercially available. With the use of highly specific antisera, the samples are assayed in duplicate following the methods recommended by the manufacturers (for assays of 6-keto-PGF$_{1\alpha}$, TxB$_2$, PGF$_{2\alpha}$, LTB$_4$, LTC$_4$ and 15-HETE, Amersham, St. Louis, MO; for assays of PGE$_2$, 5-HETE, and 12-HETE, Advanced Magnetics, Cambridge, MA). Detection limits for each kit (picograms per tube) is 14 pg of 6-keto-PGF$_{1\alpha}$, 5 pg of TxB$_2$, 8.2 pg of PGE$_2$, 3 pg of PGF$_{2\alpha}$, 1.6 pg of LTB$_4$, 8 pg of LTC$_4$, 8.2 pg of 5-HETE, 8.2 pg of 12-HETE, or 16 pg of 15-HETE.

6.2. Enzyme immunoassay

Similar to RIA, this assay is also based on the competition between an eicosanoid and a constant amount of chemiluminescent (e.g., acetylcholinesterase or alkaline phosphatase) conjugated eicosanoid for a limited amount of eicosanoid antibody that is attached to the well. The fraction corresponding to each eicosanoid is collected after purification as described previously and measured by EIA. All standards and samples should be run in duplicate. After a simultaneous incubation for all samples at 4° for overnight, the excess reagents are washed away and substrate is added to each sample. After another incubation, the enzyme reaction is stopped, and the yellow color generated is read on a microplate reader at a wavelength of 412 nm for acetylcholinesterase reaction and 405 nm for alkaline phosphatase reaction. The intensity of the bound yellow color is inversely proportional to the concentration of the eicosanoid in either standards or samples. Similar to the RIA method, percentage bound = [(average bound OD − average NSB OD)/[average maximum binding optical density

[OD] – average NSB maximum binding OD)] × 100. The maximum binding OD is obtained from the well with 0 pg/ml of standard. Each eicosanoid standard curve is obtained by plotting percentage bound versus concentration of added eicosanoid standard. The concentration of eicosanoid in biological samples is determined by interpolation. Because EIA kits for most eicosanoids are now commercially available, the detailed procedures for measurement of each eicosanoid are provided by the manufacturers. For example, there are EIA kits available for PGE_2, $PGF_{2\alpha}$, PGD_2, PGEM, 6-keto-$PGF_{1\alpha}$, 11β-$PGF_{2\alpha}$, 13,14-dihydro-15-keto-$PGF_2\alpha$, 2,3-dinor-6-keto-$PGF_1\alpha$, TXB_2, 11-dehydro TXB_2, 2.3-dinor TXB_2, LTB_4, LTB_4, LTC_4, LTE_4, and 15(S)-HETE from Cayman Chemical Co. (Ann Arbor, MI). The detection limit for each EIA kit is 10 pg/ml of PGE_2, 6 pg/ml of $PGF_{2\alpha}$, 140 pg/ml of PGD_2, 1.5 pg/ml of PGEM, 4 pg/ml of 6-keto-$PGF_{1\alpha}$, 4 pg/ml of 11β-$PGF_{2\alpha}$, 12 pg/ml of 13,14-dihydro-15-keto-$PGF_{2\alpha}$, 40 pg/ml of 2,3-dinor-6-keto-$PGF_{1\alpha}$, 22 pg/ml of TXB_2, 25 pg/ml of 11-dehydro TXB_2, 25 pg/ml of 2.3-dinor TXB_2, 10 pg/ml of LTB_4, 50 pg/ml of LTC_4, 10 pg/ml of LTE_4, or 200 pg/ml of 15(S)-HETE. In addition, Assay Designs Inc. (Ann Arbor, MI) has EIA kits for 15d-PGJ_2, 12(S)-HETE, and 13(S)-HODE. The detection limit for each EIA kit is 195 pg/ml of 15d-PGJ_2, 195 pg/ml of 12(S)-HETE, or 3900 pg/ml of 13(S)-HODE.

6.3. Gas chromatography-mass spectrometry

Electron-capture, negative-ion, chemical ionization mass spectrometry (NCI-MS) has been used for the quantitative analysis of eicosanoids in a precise, accurate, and sensitive manner. An important requirement for this assay is the availability of appropriate stable isotope analogues, which are used as internal standards. The high sensitivity of NCI-MS increases the risk of interfering substances from the biological matrix eluting at the same GC retention time as target eicosanoid. Such interference can be minimized by the use of suitable purification techniques as describe previously. All eicosanoids and their related deuterated standards can be purchased from Cayman Chemical Co. (Ann Arbor, MI). The pentafluorobenzyl (PFB) ester is the most widely used electron-capturing derivative for eicosanoids. However, eicosanoids often contain polar hydroxyl groups that cause tailing during analysis by gas chromatography (GC). This problem can be overcome by conversion of the hydroxyl groups to trimethylsilyl (TMS) ethers. TMS ethers possess excellent GC-MS characteristics and can be formed in quantitative yield. However, a ketone group in some eicosanoids can form unstable enol-TMS derivatives. This problem can be solved by the conversion of the ketone to a methoxime (MO) derivative using O-methoxylamine HCl before any derivatization.

6.3.1. Analysis of prostanoids

The dried PGs are supplemented with 25 μl deuterated PG standards (3 ng for each) and 125 μl acetone, and the combined solution is dried under a steady stream of N_2. As soon as the specimen is dried, 200 μl of 2% O-methoxylamine HCl in pyridine is added to each sample and incubated at room temperature for 30 min. The samples are stored at $-20°$ until quantification by gas chromatography-mass spectrometry. The O-methyloxime derivatives PGs are converted to pentafluorobenzyl (PFB) esters by treatment with a mixture of 40 μl of 10% (v/v) pentafluorobenzylbromide in acetonitrile and 20 μl of 10% (v/v) N,N-diisopropylethylamine in acetonitrile at room temperature for 30 min. The reaction mixtures are then dried under nitrogen, and this procedure is repeated to ensure quantitative PFB esterification. After the second esterification, the reaction mixtures are dried under N_2 and the residue is subjected to TLC using the solvent system ethyl acetate:methanol (98:2). Approximately 2 to 5 μg of the pentafluorobenzyl ester of each PG standard is subjected to TLC on a separate lane. Corresponding zones containing the prostanoids in the unknown samples are scraped and the compounds are eluted with methanol as described previously. The compound is dried under nitrogen and the residue is then converted to trimethylsilyl (TMS) ether derivatives by adding 20 μl of N,O-bis(trimethylsilyl)trifluoroacetamide (BSTFA) and 10 μl of dimethylformamide and incubating at 40° for 20 min. The reaction mixtures are then dried under nitrogen and the residue is redissolved in 10 μl of undecane.

Mass spectrometry is performed on a Nermag R10–10C mass spectrometer interfaced to an HP 5890 gas chromatograph (Hewlett-Packard). Gas chromatographic analyses are carried out on a DB1701 or DB-1 fused silica capillary column (15 m, 0.25-mm ID, 0.25-μm film thickness, J&W Scientific, Folsom, CA). One microliter of the sample is injected using the splitless mode with an injector temperature of 250°. The column temperature is programmed from 190° to 300° at 20°/min. Methane is used as the carrier gas for negative-ion chemical ionization at a flow rate of 1 ml/min. The mass spectrometer operating conditions are typically: the ion source temperature is 250°, the electron energy is 70 eV, and the filament current is 0.25 mA. Standard curves for prostanoids are prepared by addition of each prostanoid (0.1 to 10 ng) so that a constant amount of deuterium-labeled prostanoid (3 ng) is included before derivatization. The areas under the chromatographic peaks defined for selected ions are obtained from GC/MS analyses. The ratio of peak areas is plotted against the amount of each prostanoid added, which results in a linear relation. The amount of endogenous prostanoids is determined by plotting the ratio of peak areas of unknown sample against the standard curve. For example, quantification is accomplished by selected ion monitoring of the ratios of the M-PFB ions for PGE_2 (m/z 524) and [$^2H_4$] PGE_2 (m/z 528), $PGF_{1\alpha}$ (m/z 614) and [$^2H_4$] $PGF_{1\alpha}$ (m/z 618), TXB_2 (m/z 614) and [$^2H_4$] TXB_2 (m/z 618), and PGD_2 (m/z 524) and the sum of [$^2H_6$] PGD_2 (m/z 530), [$^2H_7$] PGD_2 (m/z 531), and [$^2H_8$] PGD_2 (m/z 531) (Giardiello $et\ al.$, 2004; Green $et\ al.$, 1973; Parsons and Roberts, 1988).

6.3.2. Analysis of HETEs and EETs

The samples from the extraction are esterified to the pentafluorobenzyl ester and then converted to trimethylsilyl (TMS) ether derivatives as described previously. After derivatization, the samples are dissolved in decane and analyzed by GC-MS. The column temperature is programmed from 170° to 300° at 25°/min for analysis of EETs, or from 150° to 300° at 15°/min for analysis of HETEs. Helium is used as a carrier gas with a linear velocity of 0.4 m/s. Electron–capture ionization is carried out using methane as a moderating gas at a flow resulting in ion source pressure of approximately 1.5 mm Hg at ion source temperature of 200° and electron energy of 180 eV. Under these conditions, all four EETs are eluted as a single chromatographic peak, whereas the HETE isomers (16-, 17-, 18-, 19-, and 20-HETE) are fully separated one from another. Selected ion monitoring is used to record ion abundances at m/z 319 and m/z 391 (endogenous EETs and HETEs, respectively) and m/z 327 and m/z 394 (internal standards, [$^2H_8$] 14,15-EET, and [$^2H_3$] 19-HETE, respectively). Standard curves are prepared by addition of 14,15-EET or 19-HETE (2 to 100 ng) to constant amounts of [$^2H_8$] 14,15-EET or [$^2H_3$] 19-HETE (10 or 50 ng) before derivatization, respectively (Wiswedel $et\ al.$, 2000; Zhu $et\ al.$, 1995). An identical approach is taken to quantify the amounts of EETs and HETEs as described previously.

HETEs exhibit extensive tailing and are often de-composed during GC separation. These problems can be overcome by hydrogenation of the HETEs prior to GC-MS analysis. Lipids from biological samples are immediately hydrogenated by treatment with hydrogen gas using PtO_2 as catalyst in methanol/ethylacetate (1:1) followed by alkaline hydrolysis (KOH/ethanol, 30 min, 40°). The hydrolysis mixture is acidified to pH 1.0 and the hydroxy fatty acids are extracted by a solid–phase extraction as described previously. After derivatization by pentafluorobenzylbromide and bis-(trimethylsilyl)trifluoroacetamide, samples are analyzed by GC-MS as described previously. For example, the mass spectrum of hydrogenated 12(S)-HETE-TMS-PFB and its corresponding internal standard [$^2H_3$] 12 (S)-HETE-TMS-PFB shows major ions at m/z 399 (M-PFB) and 422 (M-PFB), respectively. The level of each HETE isomer is determined based on the inclusion of known quantities of deuterated HETE isomer as internal standards.

6.3.3. Analysis of LTB$_4$

LTB$_4$ is esterified to the PFB ester and then converted to TMS ether derivatives as described previously. Each sample (1.0 μl) is injected with the oven temperature set at 50°. After 1 min the temperature is increased to 265° at 20°/min and then programmed to 290° at 5°/min. The mass spectrometer operating conditions are typically accelerating voltage, 8 kV; ionization energy, 80 eV; emission current, 1 mA; and resolution, 1000. Methane is used as the carrier gas at a pressure of 1×10^{-5} torr measured at the source housing. LTB$_4$ is quantified by monitoring the (M-PFB) ion at m/z 479 for LTB$_4$ and the corresponding ion at m/z 483 for the [^{2}H$_4$] LTB$_4$ as an internal standard using the selected-ion–monitoring scan mode (Mathews 1990). A standard curve is constructed by adding a constant amount of [^{2}H$_4$] LTB4 (10 ng, 2.9 pmol) to 0–100 ng (0–298 pmol) LTB$_4$ followed by derivatization as described previously. The amount of LTB$_4$ in the unknown samples is determined as described previously.

6.3.4. Liquid chromatography-mass spectrometry-mass spectrometry

Analysis of eicosanoids by GC-MS method is time consuming and complicated. Recently, LC-MS-MS has been used to simultaneously quantify endogenous eicosanoids in *in vitro* and *in vivo* samples without purification procedures (Kempen *et al.*, 2001; Yang *et al.*, 2002, 2006).

Before extraction, each sample collected from cell-free tissue culture medium, urine, plasma, or tissue homogenate is mixed with a mixture of known quantities of deuterated eicosanoids of interest (2 ng for each deuterated eicosanoid). The samples are extracted by organic solvent as described previously, and the dried extract is reconstituted in 100 μl of methanol:ammonium acetate buffer (10 mM at pH 8.5, 70:30). Standard curves are prepared for each of eicosanoid of interest at concentrations ranging from 1 to 100 ng/ml. LC-MS-MS analyses are performed on a Micromass Quattro Ultima triple quadrupole analyzer (Macromass, Beverly, MA) equipped with an Agilent HP 1100 binary pump HPLC inlet. Eicosanoids are separated using a Luna 3μ Phenyl-Hexyl 2 × 150 mm LC column (Phenomenex, Torrance, CA). Twenty-five microliters of sample are injected into the column. The mobile phase consists of 10-mM ammonium acetate (pH 8.5) and methanol. Due to structural similarity among different PGs, a linear methanol gradient from 40 to 60% for 18 min and then from 60 to 100% in 1 min and at 100% methanol for 4 min is used to achieve chromatographic baseline resolution for the PGs. For the analysis of HETEs, LTB$_4$, and 13-HODE, the separation is achieved using a rapid linear gradient of 70 to 90% methanol with injection-to-injection time of 10 min. The flow rate is 250 μl/min with a column temperature of 50°. The mass spectrometer is operated in the electrospray negative-ion mode with a cone voltage of 100 V, a cone gas flow rate of 117 l/h, and a devolution gas flow rate of 998 l/h. The temperature of the desolvation region is 400°, and the temperature of the source region is 120°. Fragmentation for all compounds is performed using argon as the collision gas at a collision cell pressure of 2.10×10^{-3} torr. The collision energy is 19 V. All eicosanoids are detected using electrospray negative-ionization and multiple-reaction monitoring of the transition ions for the metabolites and their internal standards.

For analysis of the major urinary metabolite of PGE_2, 13,14-dihydro-15-keto-PGA_2 (PGEM), 400 μl of urine is added with 12.4 ng of $[^2H_6]$ PGE-M as internal standard and derivatized using methoximine HCl (16% w/v) in 1.5 M of sodium acetate solution. Samples are then extracted by a C18 SepPak and dried under N_2 as described previously. The dry extracts are resuspended in 50-μl mobile phase A (5-mM ammonium acetate:aceto-nitrile:acetic acid, 95:4.9:0.1) and analyzed by LC–MS/MS as described previously (Mann *et al.*, 2006).

ACKNOWLEDGMENTS

This work is supported, in part, by the National Institutes of Health (NIH) (grants RO1DK 62112, P01-CA-77839, R37-DK47297, and P30 CA068485) (R. N. D.). R. N. D. is the B. F. Byrd Professor of Molecular Oncology at Vanderbilt University, and is the recipient of an NIH MERIT award (R37-DK47297). We also thank the T. J. Martell Foundation and the National Colorectal Cancer Research Alliance (NCCRA) for generous support (R. N. D.).

REFERENCES

Adderley, S. R., and Fitzgerald, D. J. (1999). Oxidative damage of cardiomyocytes is limited by extracellular regulated kinases 1/2–mediated induction of cyclooxygenase-2. *J. Biol. Chem.* **274**, 5038–5046.

Amin, A. R., Attur, M., and Abramson, S. B. (1999). Nitric oxide synthase and cyclooxygenases: Distribution, regulation, and intervention in arthritis. *Curr. Opin. Rheumatol.* **11**, 202–209.

Amin, M. (1975). Direct quantitative thin-layer chromatographic determination of prostaglandins A2, B2, E2 and F2 alpha. *J. Chromatogr.* **108**, 313–321.

Anderson, G. D., Hauser, S. D., McGarity, K. L., Bremer, M. E., Isakson, P. C., and Gregory, S. A. (1996). Selective inhibition of cyclooxygenase (COX)-2 reverses inflammation and expression of COX-2 and interleukin 6 in rat adjuvant arthritis. *J. Clin. Invest.* **97**, 2672–2679.

Beneytout, J. L., Greuet, D., Tixier, M., and Rigaud, M. (2005). Separation of arachidonic acid metabolites by thin-layer chromatography using new silicone-bonded plates. *J. High Res. Chromatogr. Chromatogr. Commun.* **7**, 538–539.

Bhatia, B., Maldonado, C. J., Tang, S., Chandra, D., Klein, R. D., Chopra, D., Shappell, S. B., Yang, P., Newman, R. A., and Tang, D. G. (2003). Subcellular localization and tumor-suppressive functions of 15-lipoxygenase 2 (15-LOX2) and its splice variants. *J. Biol. Chem.* **278**, 25091–25100.

Bortuzzo, C., Hanif, R., Kashfi, K., Staiano-Coico, L., Shiff, S. J., and Rigas, B. (1996). The effect of leukotrienes B and selected HETEs on the proliferation of colon cancer cells. *Biochim. Biophys. Acta* **1300,** 240–246.

Brouard, C., and Pascaud, M. (1990). Effects of moderate dietary supplementations with n-3 fatty acids on macrophage and lymphocyte phospholipids and macrophage eicosanoid synthesis in the rat. *Biochim. Biophys. Acta* **1047,** 19–28.

Bygdeman, M., and Samuelsson, B. (1964). Quantitative determination of prostaglandins in human semen. *Clin. Chim. Acta* **10,** 566–568.

Campbell, W. B., and Ojeda, S. R. (1987). Measurement of prostaglandins by radioimmunoassay. *Methods Enzymol.* **141,** 323–341.

Capdevila, J. H., Falck, J. R., and Harris, R. C. (2000). Cytochrome P450 and arachidonic acid bioactivation. Molecular and functional properties of the arachidonate monooxygenase. *J. Lipid Res.* **41,** 163–181.

Catella, F., Nowak, J., and Fitzgerald, G. A. (1986). Measurement of renal and non-renal eicosanoid synthesis. *Am. J. Med.* **81,** 23–29.

Cellai, I., Benvenuti, S., Luciani, P., Galli, A., Ceni, E., Simi, L., Baglioni, S., Muratori, M., Ottanelli, B., Serio, M., Thiele, C. J., and Peri, A. (2006). Antineoplastic effects of rosiglitazone and PPARgamma transactivation in neuroblastoma cells. *Br. J. Cancer* **95,** 879–888.

Chen, J. K., Falck, J. R., Reddy, K. M., Capdevila, J., and Harris, R. C. (1998). Epoxyeicosatrienoic acids and their sulfonimide derivatives stimulate tyrosine phosphorylation and induce mitogenesis in renal epithelial cells. *J. Biol. Chem.* **273,** 29254–29261.

Chen, J. K., Wang, D. W., Falck, J. R., Capdevila, J., and Harris, R. C. (1999). Transfection of an active cytochrome P450 arachidonic acid epoxygenase indicates that 14,15-epoxyeicosatrienoic acid functions as an intracellular second messenger in response to epidermal growth factor. *J. Biol. Chem.* **274,** 4764–4769.

Craven, P. A., and DeRubertis, F. R. (1983). Patterns of prostaglandin synthesis and degradation in isolated superficial and proliferative colonic epithelial cells compared to residual colon. *Prostaglandins* **26,** 583–604.

Del Vecchio, R. P., Maxey, K. M., and Lewis, G. S. (1992). A quantitative solid-phase enzymeimmunoassay for 13,14-dihydro-15-keto-prostaglandin F2 alpha in plasma. *Prostaglandins* **43,** 321–330.

DuBois, R. N., Abramson, S. B., Crofford, L., R.A., G.,Simon,L. S., Van De Putte, L. B., and Lipsky, P. E. (1998). Cyclooxygenase in biology and disease. *FASEB J.* **12,** 1063–1073.

Eastman, A. R., and Dowsett, M. (1976). The simultaneous separation of individual prostaglandins by thin-layer chromatography on an unmodified support. *J. Chromatogr.* **128,** 224–226.

Eling, T., Tainer, B., Ally, A., and Warnock, R. (1982). Separation of arachidonic acid metabolites by high-pressure liquid chromatography. *Methods Enzymol.* **86,** 511–517.

Ensor, C. M., Yang, J. Y., Okita, R. T., and Tai, H. H. (1990). Cloning and sequence analysis of the cDNA for human placental NAD(+)-dependent 15-hydroxyprostaglandin dehydrogenase. *J. Biol. Chem.* **265,** 14888–14891.

Fleming, I., Michaelis, U. R., Bredenkotter, D., Fisslthaler, B., Dehghani, F., Brandes, R. P., and Busse, R. (2001). Endothelium-derived hyperpolarizing factor synthase (cytochrome P450 2C9) is a functionally significant source of reactive oxygen species in coronary arteries. *Circ. Res.* **88,** 44–51.

Forman, B. M., Chen, J., and Evans, R. M. (1997). Hypolipidemic drugs, polyunsaturated fatty acids, and eicosanoids are ligands for peroxisome proliferator-activated receptors alpha and delta. *Proc. Natl. Acad. Sci. USA* **94,** 4312–4317.

Forman, B. M., Tontonoz, P., Chen, J., Brun, R. P., Spiegelman, B. M., and Evans, R. M. (1995). 15-Deoxy-delta 12, 14-prostaglandin J2 is a ligand for the adipocyte determination factor PPAR gamma. *Cell* **83,** 803–812.

Frolich, J. C. (1984). Measurement of icosanoids. Report of the Group for Standardization of Methods in Icosanoid Research. *Prostaglandins* **27**, 349–368.

Giardiello, F. M., Casero, R. A., Jr., Hamilton, S. R., Hylind, L. M., Trimbath, J. D., Geiman, D. E., Judge, K. R., Hubbard, W., Offerhaus, G. J., and Yang, V. W. (2004). Prostanoids, ornithine decarboxylase, and polyamines in primary chemoprevention of familial adenomatous polyposis. *Gastroenterology* **126**, 425–431.

Gould, S. R., Brash, A. R., Conolly, M. E., and Lennard-Jones, J. E. (1981). Studies of prostaglandins and sulphasalazine in ulcerative colitis. *Prostaglandins Med.* **6**, 165–182.

Green, K., Granstrom, E., Samuelsson, B., and Axen, U. (1973). Methods for quantitative analysis of PGF2, PGE2, 9, 11 -dihydroxy-15-keto-prost-5-enoic acid and 9, 11, 15-trihydroxy-prost-5-enoic acid from body fluids using deuterated carriers and gas chromatography-mass spectrometry. *Anal. Biochem.* **54**, 434–453.

Green, K., and Samuelsson, B. (1964). Prostaglandins and related factors. XIX. Thin-layer chromatography of prostaglandins. *J. Lipid Res.* **15**, 117–120.

Gupta, R. A., Tan, J., Krause, W. F., Geraci, M. W., Willson, T. M., Dey, S. K., and DuBois, R. N. (2000). Prostacyclin-mediated activation of peroxisome proliferator-activated receptor delta in colorectal cancer. *Proc. Natl. Acad. Sci. USA* **97**, 13275–13280.

Gupta, R. A., Wang, D., Katkuri, S., Wang, H., Dey, S. K., and DuBois, R. N. (2004). Activation of nuclear hormone receptor peroxisome proliferator-activated receptor-delta accelerates intestinal adenoma growth. *Nat. Med.* **10**, 245–247.

Hansen-Petrik, M. B., McEntee, M. F., Jull, B., Shi, H., Zemel, M. B., and Whelan, J. (2002). Prostaglandin E(2) protects intestinal tumors from nonsteroidal anti-inflammatory drug-induced regression in Apc(Min/+) mice. *Cancer Res.* **62**, 403–408.

Herschman, H. R. (1996). Prostaglandin synthase 2. *Biochem. Biophys. Acta* **1299**, 125–140.

Herschman, H. R., Xie, W., and Reddy, S. (1995). Inflammation, reproduction, cancer and all that... The regulation and role of the inducible prostaglandin synthase. *Bioessays* **17**, 1031–1037.

Hoebel, B. G., and Graier, W. F. (1998). 11,12-Epoxyeicosatrienoic acid stimulates tyrosine kinase activity in porcine aortic endothelial cells. *Eur. J. Pharmacol.* **346**, 115–117.

Hollenberg, S. M., Tong, W., Shelhamer, J. H., Lawrence, M., and Cunnion, R. E. (1994). Eicosanoid production by human aortic endothelial cells in response to endothelin. *Am. J. Physiol.* **267**, H2290–H2296.

Hussey, H. J., Bibby, M. C., and Tisdale, M. J. (1996). Novel anti-tumour activity of 2,3,5-trimethyl-6-(3-pyridylmethyl)-1,4-benzoquinone (CV-6504) against established murine adenocarcinomas (MAC). *Br. J. Cancer* **73**, 1187–1192.

Hussey, H. J., and Tisdale, M. J. (1994). Effect of polyunsaturated fatty acids on the growth of murine colon adenocarcinomas *in vitro* and *in vivo*. *Br. J. Cancer* **70**, 6–10.

Jaffe, B. M., Behrman, H. R., and Parker, C. W. (1973). Radioimmunoassay measurement of prostaglandins E, A, and F in human plasma. *J. Clin. Invest.* **52**, 398–405.

Kawamori, T., Uchiya, N., Sugimura, T., and Wakabayashi, K. (2003). Enhancement of colon carcinogenesis by prostaglandin E2 administration. *Carcinogenesis* **24**, 985–990.

Keirse, M. J., and Turnbull, A. C. (1975). Metabolism of prostaglandins within the pregnant uterus. *Br. J. Obstet. Gynaecol.* **82**, 887–893.

Kempen, E. C., Yang, P., Felix, E., Madden, T., and Newman, R. A. (2001). Simultaneous quantification of arachidonic acid metabolites in cultured tumor cells using high-performance liquid chromatography/electrospray ionization tandem mass spectrometry. *Anal. Biochem.* **297**, 183–190.

Kliewer, S. A., Lenhard, J. M., Willson, T. M., Patel, I., Morris, D. C., and Lehmann, J. M. (1995). A prostaglandin J2 metabolite binds peroxisome proliferator-activated receptor gamma and promotes adipocyte differentiation. *Cell* **83**, 813–819.

Lawson, L. D., and Powell, D. W. (1987). Bradykinin-stimulated eicosanoid synthesis and secretion by rabbit ileal components. *Am. J. Physiol.* **252**, G783–G790.

LeDuc, L. E., and Needleman, P. (1980). Prostaglandin synthesis by dog gastrointestinal tract. *Adv. Prostaglandin Thromboxane Res.* **8,** 1515–1517.

Lee, D. Y., Lupton, J. R., and Chapkin, R. S. (1992). Prostaglandin profile and synthetic capacity of the colon: Comparison of tissue sources and subcellular fractions. *Prostaglandins* **43,** 143–164.

Liston, T. E., and Roberts, L. J., 2nd. (1985). Transformation of prostaglandin D2 to 9 alpha, 11 beta-(15S)-trihydroxyprosta-(5Z,13E)-dien-1-oic acid (9 alpha, 11 beta–prostaglandin F2): A unique biologically active prostaglandin produced enzymatically *in vivo* in humans. *Proc. Natl. Acad. Sci. USA* **82,** 6030–6034.

MacDermott, R. P. (1994). Alterations in the mucosal immune system in ulcerative colitis and Crohn's disease. *Med. Clin. North Am.* **78,** 1207–1231.

Mann, J. R., Backlund, M. G., Buchanan, F. G., Daikoku, T., Holla, V. R., Rosenberg, D. W., Dey, S. K., and DuBois, R. N. (2006). Repression of prostaglandin dehydrogenase by epidermal growth factor and snail increases prostaglandin E2 and promotes cancer progression. *Cancer Res.* **66,** 6649–6656.

Mathews, W. R. (1990). Quantitative gas chromatography-mass spectrometry analysis of leukotriene B4. *Methods Enzymol.* **187,** 76–81.

McAdam, B. F., Catella-Lawson, F., Mardini, I. A., Kapoor, S., Lawson, J. A., and FitzGerald, G. A. (1999). Systemic biosynthesis of prostacyclin by cyclooxygenase (COX)-2: The human pharmacology of a selective inhibitor of COX-2. *Proc. Natl. Acad. Sci. USA* **96,** 272–277.

Meyer, H. H., Eisele, K., and Osaso, J. (1989). A biotin–streptavidin amplified enzymeimmunoassay for 13,14–dihydro–15–keto-PGF2 alpha. *Prostaglandins* **38,** 375–383.

Mutoh, M., Watanabe, K., Kitamura, T., Shoji, Y., Takahashi, M., Kawamori, T., Tani, K., Kobayashi, M., Maruyama, T., Kobayashi, K., Ohuchida, S., Sugimoto, Y., *et al.* (2002). Involvement of prostaglandin E receptor subtype EP(4) in colon carcinogenesis. *Cancer Res.* **62,** 28–32.

Needleman, P., and Isakson, P. C. (1997). The discovery and function of COX-2. *J. Rheumatol.* **24**(Suppl. 49), 6–8.

O'Sullivan, S., Dahlen, B., Dahlen, S. E., and Kumlin, M. (1996). Increased urinary excretion of the prostaglandin D2 metabolite 9 alpha, 11 beta–prostaglandin F2 after aspirin challenge supports mast cell activation in aspirin-induced airway obstruction. *J. Allergy Clin. Immunol.* **98,** 421–432.

O'Sullivan, S., Mueller, M. J., Dahlen, S. E., and Kumlin, M. (1999). Analyses of prostaglandin D2 metabolites in urine: Comparison between enzyme immunoassay and negative ion chemical ionisation gas chromatography-mass spectrometry. *Prostaglandins Other Lipid Mediat.* **57,** 149–165.

Ohm, K., Albers, H. K., and Lisboa, B. P. (1984). Measurement of eight prostaglandins in human gingival and periodontal disease using high pressure liquid chromatography and radioimmunoassay. *J. Periodontal Res.* **19,** 501–511.

Panigrahy, D., Huang, S., Kieran, M. W., and Kaipainen, A. (2005). PPARgamma as a therapeutic target for tumor angiogenesis and metastasis. *Cancer Biol. Ther.* **4,** 687–693.

Parsons, W. G., 3rd, and Roberts, L. J., 2nd. (1988). Transformation of prostaglandin D2 to isomeric prostaglandin F2 compounds by human eosinophils. A potential mast cell–eosinophil interaction. *J. Immunol.* **141,** 2413–2419.

Patterson, L. H., and Murray, G. I. (2002). Tumour cytochrome P450 and drug activation. *Curr. Pharm. Des.* **8,** 1335–1347.

Portanova, J. P., Zhang, Y., Anderson, G. D., Hauser, S. D., Masferrer, J. L., Seibert, K., Gregory, S. A., and Isakson, P. C. (1996). Selective neutralization of prostaglandin E2 blocks inflammation, hyperalgesia, and interleukin 6 production *in vivo*. *J. Exp. Med* **184,** 883–891.

Potente, M., Fisslthaler, B., Busse, R., and Fleming, I. (2003). 11,12-Epoxyeicosatrienoic acid-induced inhibition of FOXO factors promotes endothelial proliferation by down-regulating p27Kip1. *J. Biol. Chem.* **278**, 29619–29625.

Powell, W. S. (1980). Rapid extraction of oxygenated metabolites of arachidonic acid from biological samples using octadecylsilyl silica. *Prostaglandins* **20**, 947–957.

Rigas, B., Goldman, I. S., and Levine, L. (1993). Altered eicosanoid levels in human colon cancer. *J. Lab. Clin. Med.* **122**, 518–523.

Roberts, L. J., 2nd, Sweetman, B. J., and Oates, J. A. (1981). Metabolism of thromboxane B2 in man. Identification of twenty urinary metabolites. *J. Biol. Chem.* **256**, 8384–8393.

Rosenkranz, B., Fischer, C., Reimann, I., Weimer, K. E., Beck, G., and Frölich, J. C. (1980). Identification of the major metabolite of prostacyclin and 6-ketoprostaglandin F1 alpha in man. *Biochim. Biophys. Acta* **619**, 207–213.

Salmon, J. A. (1983). Measurement of eicosanoids by bioassay and radioimmunoassay. *Br. Med. Bull.* **39**, 227–231.

Samuelsson, B., Goldyne, M., Granstrom, E., Hamberg, M., Hammarstrom, S., and Malmsten, C. (1978). Prostaglandins and thromboxanes. *Annu. Rev. Biochem.* **47**, 997–1029.

Samuelsson, B., Granstrom, E., Green, K., Hamberg, M., and Hammarstrom, S. (1975). Prostaglandins. *Annu. Rev. Biochem.* **44**, 669–695.

Sanduja, S. K., Mehta, K., Xu, X. M., Hsu, S. M., Sanduja, R., and Wu, K. K. (1991). Differentiation-associated expression of prostaglandin H and thromboxane A synthases in monocytoid leukemia cell lines. *Blood* **78**, 3178–3185.

Shaw, J. E., and Ramwell, P. W. (1969). Separation, identification, and estimation of prostaglandins. *Methods Biochem. Anal.* **17**, 325–371.

Shureiqi, I., and Lippman, S. M. (2001). Lipoxygenase modulation to reverse carcinogenesis. *Cancer Res.* **61**, 6307–6312.

Shureiqi, I., Wojno, K. J., Poore, J. A., Reddy, R. G., Moussalli, M. J., Spindler, S. A., Greenson, J. K., Normolle, D., Hasan, A. A., Lawrence, T. S., and Brenner, D. E. (1999). Decreased 13-S-hydroxyoctadecadienoic acid levels and 15-lipoxygenase-1 expression in human colon cancers. *Carcinogenesis* **20**, 1985–1995.

Shureiqi, I., Wu, Y., Chen, D., Yang, X. L., Guan, B., Morris, J. S., Yang, P., Newman, R. A., Broaddus, R., Hamilton, S. R., Lynch, P., Levin, B., *et al.* (2005). The critical role of 15-lipoxygenase-1 in colorectal epithelial cell terminal differentiation and tumorigenesis. *Cancer Res.* **65**, 11486–11492.

Smith, G. S., Warhurst, G., and Turnberg, L. A. (1982). Synthesis and degradation of prostaglandin E2 in the epithelial and sub-epithelial layers of the rat intestine. *Biochim. Biophys. Acta* **713**, 684–687.

Smith, W. L., DeWitt, D. L., and Garavito, R. M. (2000). Cyclooxygenases: Structural, cellular, and molecular biology. *Annu. Rev. Biochem.* **69**, 145–182.

Sonoshita, M., Takaku, K., Sasaki, N., Sugimoto, Y., Ushikubi, F., Narumiya, S., Oshima, M., and Taketo, M. M. (2001). Acceleration of intestinal polyposis through prostaglandin receptor EP2 in Apc(Delta 716) knockout mice. *Nat. Med.* **7**, 1048–1051.

Spector, A. A., Fang, X., Snyder, G. D., and Weintraub, N. L. (2004). Epoxyeicosatrienoic acids (EETs): Metabolism and biochemical function. *Prog. Lipid Res.* **43**, 55–90.

Takayama, O., Yamamoto, H., Damdinsuren, B., Sugita, Y., Ngan, C. Y., Xu, X., Tsujino, T., Takemasa, I., Ikeda, M., Sekimoto, M., Matsuura, N., and Monden, M. (2006). Expression of PPARdelta in multistage carcinogenesis of the colorectum: Implications of malignant cancer morphology. *Br. J. Cancer* **95**, 889–895.

Thompson, C. J., Los, M., and Horton, E. W. (1970). The separation, identification and estimation of prostaglandins in nanogram quantities by combined gas chromatography-mass spectrometry. *Life Sci. I* **9**, 983–988.

Wang, D., Mann, J. R., and DuBois, R. N. (2005). The role of prostaglandins and other eicosanoids in the gastrointestinal tract. *Gastroenterology* **128,** 1445–1461.

Wang, D., Wang, H., Shi, Q., Katkuri, S., Walhi, W., Desvergne, B., Das, S. K., Dey, S. K., and DuBois, R. N. (2004). Prostaglandin E(2) promotes colorectal adenoma growth via transactivation of the nuclear peroxisome proliferator-activated receptor delta. *Cancer Cell* **6,** 285–295.

Watanabe, K., Kawamori, T., Nakatsugi, S., Ohta, T., Ohuchida, S., Yamamoto, H., Maruyama, T., Kondo, K., Ushikubi, F., Narumiya, S., Sugimura, T., and Wakabayashi, K. (1999). Role of the prostaglandin E receptor subtype EP1 in colon carcinogenesis. *Cancer Res.* **59,** 5093–5096.

Wiswedel, I., Bohne, M., Hirsch, D., Kuhn, H., Augustin, W., and Gollnick, H. (2000). A sensitive gas chromatography-mass spectrometry assay reveals increased levels of mono-hydroxyeicosatetraenoic acid isomers in human plasma after extracorporeal photoimmunotherapy and under *in vitro* ultraviolet A exposure. *J. Invest. Dermatol.* **115,** 499–503.

Yang, P., Chan, D., Felix, E., Madden, T., Klein, R. D., Shureiqi, I., Chen, X., Dannenberg, A. J., and Newman, R. A. (2006). Determination of endogenous tissue inflammation profiles by LC/MS/MS: COX- and LOX-derived bioactive lipids. *Prostaglandins Leukot. Essent. Fatty Acids* **75,** 385–395.

Yang, P., Felix, E., Madden, T., Fischer, S. M., and Newman, R. A. (2002). Quantitative high-performance liquid chromatography/electrospray ionization tandem mass spectrometric analysis of 2- and 3-series prostaglandins in cultured tumor cells. *Anal. Biochem.* **308,** 168–177.

Ye, Y. N., Wu, W. K., Shin, V. Y., Bruce, I. C., Wong, B. C., and Cho, C. H. (2005). Dual inhibition of 5-LOX and COX-2 suppresses colon cancer formation promoted by cigarette smoke. *Carcinogenesis* **26,** 827–834.

Yin, Y., Russell, R. G., Dettin, L. E., Bai, R., Wei, Z. L., Kozikowski, A. P., Kopleovich, L., and Glazer, R. I. (2005). Peroxisome proliferator-activated receptor delta and gamma agonists differentially alter tumor differentiation and progression during mammary carcinogenesis. *Cancer Res.* **65,** 3950–3957.

Zhu, Y., Schieber, E. B., McGiff, J. C., and Balazy, M. (1995). Identification of arachidonate P-450 metabolites in human platelet phospholipids. *Hypertension* **25,** 854–859.

CHAPTER THREE

Noninvasive Assessment of the Role of Cyclooxygenases in Cardiovascular Health: A Detailed HPLC/MS/MS Method

Wen-Liang Song, John A. Lawson, Miao Wang, Helen Zou, *and* Garret A. FitzGerald

Contents

Abstract

A robust method for the routine quantitation of a selected group of urinary eicosanoid metabolites of interest to cardiovascular research in human

Institute for Translational Medicine and Therapeutics, School of Medicine, University of Pennsylvania, Philadelphia, Pennsylvania

Methods in Enzymology, Volume 433
ISSN 0076-6879, DOI: 10.1016/S0076-6879(07)33003-6

and mouse is described and discussed. Included are the addition of stable isotope-labeled internal standards, solid phase extraction, and quantitation by liquid chromatography/tandem mass spectrometry using selected reaction monitoring (SRM) techniques.

1. INTRODUCTION

The identification and deletion of G protein coupled receptors with high affinity for prostaglandins (PGs) and thromboxane (Tx) A_2 has revealed a remarkably diverse and often conflicting biology of these cyclooxygenase products of arachidonic acid. This suggests that the therapeutic possibilities to be gleaned from this pathway remain to be fully realized. The cardioprotective properties of low-dose aspirin (Reilly and FitzGerald, 2002) are explained sufficiently by its sustained inhibition of platelet COX-1 derived TxA_2; similarly, the gastrointestinal adverse effects of even low doses (Patrono *et al.*, 2005) are explicable in terms of this property combined with blockade of COX-1–derived prostacyclin (PGI_2) and PGE_2 in gastroduodenal epithelium, where the prostanoids afford cytoprotection. Similarly, the suppression of pain and inflammation by nonsteroidal antiinflammatory drugs (NSAIDs) is largely attributable to suppression of COX-2 derived PGI_2 and PGE_2, while the cardiovascular hazard revealed by placebo controlled trials of NSAIDs specific for COX-2 is explicable by suppression of the same products in the cardiovascular system (Grosser *et al.*, 2006). Currently, drugs that inhibit synthases downstream of the COX enzymes and PG receptor agonists and antagonists are under clinical development. The selection of rational target populations depends on identification of patient subsets in which biosynthesis (or action) of the relevant COX product is altered. For example, the incorrect conclusion that low-dose aspirin did not afford cardioprotection was drawn from a decade's worth of clinical trials in populations, which, in retrospect, were dilute with respect to patients with elevated Tx biosynthesis. When trials were performed in unstable angina, a condition in which phasic ischemic episodes are associated with major changes in Tx formation, aspirin reduced both the incidence of myocardial infarction and death by 50% (Reilly and FitzGerald, 2002). More recently, the existence of PG isomers has been characterized (Morrow *et al.*, 1992) and such isoprostanes have emerged as the "gold standard" indices of lipid peroxidation *in vivo* (Lawson *et al.*, 1999). Perhaps analogous to the experience with aspirin we presently have little evidence from clinical trials of "antioxidant" vitamins that they are effective in diseases in which it seems that oxidant stress would be likely to play a role. However, none of these trials documented biochemically that the populations exhibited increased levels of oxidant stress *in vivo*—as might be reflected by isoprostane generation—or that the vitamins were actually effectively antioxidant in the populations under study (Greindling and FitzGerald, 2003).

These examples highlight the necessity of characterizing actual biosynthesis of COX products as a step toward understanding pathophysiology, as a guide to drug development, and in the interpretation of clinical trials.

This chapter describes methods for the simultaneous measurement of metabolites of PGD_2, PGE_2, PGI_2, and TxA_2, as well as $8,12\text{-}iso\text{-}iPF_{2\alpha}\text{-}VI$, an isoprostane marker of arachidonate autooxidation. It is targeted at researchers desiring to start a new eicosanoid analysis lab or mass spectrometrists moving into the eicosanoid field. The methods presented have evolved over time. They have been desinged to be routine and take advantage of instrumentation that offers the combination of sensitivity, specificity, reliability, operational simplicity, and affordability—features necessary for the routine quantitation of picogram quantities of eicosanoid metabolites from a medium as complex and varied as urine. We will document them in detail, accompanied with minimal annotations of history and logic, so that suitably equipped labs may adopt and adapt them for their own research goals. Although the methods are for specific analytes, the principles involved can be applied to other eicosanoids.

2. BACKGROUND AND HISTORY

Since von Euler's discovery of PGs in 1934, there has been a steady growth of interest in the role of PGs and other eicosanoids in physiology and medicine, which, in turn has driven the need for their quantitation. This need has been satisfied by two major approaches: immunologic methods (RIA, EIA) and mass spectrometric methods, each of which has advantages and disadvantages. For the purposes of this chapter, we will limit comment to the general statement that immunologic methods are less expensive and less technologically demanding than mass spectrometric methods, but may be less specific, may be affected to a greater degree by losses incurred during sample preparation, and are not amenable to multiple analytes. Also beyond the scope of this chapter is a comparison of the several types of mass spectrometers that vary according to the method of separating ions of different masses. We have concluded that quadrupole mass filters, which separate on the basis of the stability of the trajectory of ions traversing a variably oscillating electronic field, offer the best combination of sensitivity, specificity, reliability, operational simplicity, and affordability.

Only two methods of sample introduction into mass spectrometers have been used extensively—gas chromatography (GC) and high-performance liquid chromatography (HPLC). GC was the first to evolve, since the low volume of volatile carrier gases, typically helium, could be efficiently evacuated by the pumping system of the mass spectrometer, which requires a high vacuum to operate. Ionization of the GC effluent typically occurred by electron impact (EI), which yields the classic mass spectrum or, later, by chemical ionization (CI), a less energetic method in which ion-molecule reactions ionize the analyte with less fragmentation. A derivative of CI is electron capture/negative ionization (EC/NI), first described in 1976 (Hunt *et al.*, 1976). Here, thermal (low-energy) electrons formed in the CI source are captured by an electrophilic group on the analyte. For eicosanoids, the pentafluorobenzyl (PFB) ester is often used because it is easily formed, is highly electrophilic, and is a good leaving group. EC/NI of the PFB ester causes very little fragmentation, and the low background of negative ions results in a high signal-to-noise ratio. This has long been the method of choice for eicosanoid analysis (Blair *et al.*, 1982; Min *et al.*, 1980), and is still the most sensitive method widely available. The many drawbacks of the method, including the necessity for extensive sample purification and required derivatization of analytes to thermally stable, volatile forms were tolerated until a practical means of interfacing an HPLC, atmospheric pressure ionization (API), was developed.

Atmospheric pressure ionization exists in two forms. In electrospray (ES), ions are formed by desolvation of a fine spray of droplets charged by exposure to a high potential field. The other form of API, atmospheric pressure chemical ionization (APCI), is suitable for compounds that do not exist as ions in solution. In APCI, molecules are indirectly ionized by reactions with ionized source gases formed by a corona discharge needle. An EC derivative of APCI, electron-capture atmospheric-pressure chemical ionization (EC APCI) is analogous to that of GC/EC/NI (Singh *et al.*, 2000), and has been used for eicosanoid quantitation. Although EC APCI is more sensitive than ES (Lee *et al.*, 2003), we decided to use ES for high-throughput routine samples because current instrumentation has enough sensitivity to take advantage of the simpler sample preparation protocol.

3. PRINCIPLES OF EICOSANOID ANALYSIS BY MASS SPECTROMETRY

The major factors to consider in developing a successful analytical method will be considered individually.

3.1. Internal standard

The choice of an appropriate internal standard (IS) is crucial. It must have characteristics as similar as possible to the analyte during extraction, chromatography, and mass spectrometry. A near-perfect IS can be obtained by labeling the analyte with stable isotopes.

Stable isotopes are nonradioactive, differing only in the number of neutrons in the atomic nucleus. Examples relevant to eicosanoid quantitation are [^{2}H] (deuterium, d), which contains two neutrons instead of the usual complement of one in the most common isotope of hydrogen [^{1}H], and [^{18}O], which has two more neutrons than the more common oxygen isotope [^{16}O]. The analogous carbon isotope, [^{13}C], is not used as often, mainly due to the fact that it is more difficult to incorporate chemically. The added neutrons change the chemical and physical properties of an eicosanoid to an almost imperceptible degree (Fig. 3.1B). When a homologous stable isotope–labeled internal standard is added to a sample and allowed to reach equilibrium with the endogenous analyte, the IS/analyte ratio is fixed; any degradation or loss upon extraction will affect both compounds in parallel, and will not affect the initial ratio. Under most circumstances, linearity can be assumed, and the IS/analyte ratio measured by the MS is that of the sample when spiked.

The optimum number of added neutrons is approximately four, as determined by the natural [^{13}C] content of the analyte. Approximately

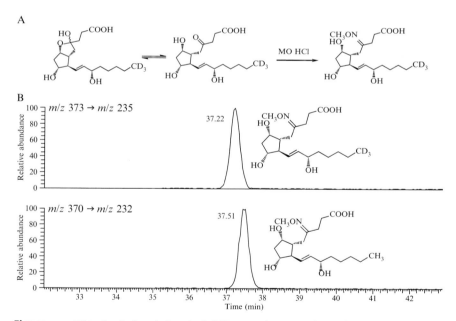

Figure 3.1 The physical and chemical differences between the endogenous 2, 3-dinor-6-keto-PGF$_{1\alpha}$ and the internal standard arising from the difference of three neutrons are minute. (A) The three extra neutrons on the IS do not affect the hemiketal equilibrium or the rate or degree of derivatization. (B) Authentic 2, 3-dinor-6-keto-PGF$_{1\alpha}$ and the IS exhibit minor differences in their interactions with the stationary phase. IS, internal standard.

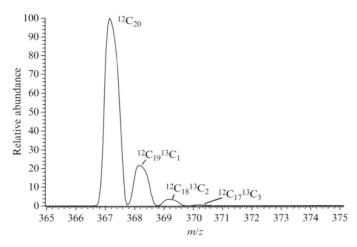

Figure 3.2 MS scan of the molecular anion region of 11-dehydro-TxB$_2$ showing the contribution of ^{13}C.

1.1% of all carbons are [^{13}C]), so the odds of an eicosanoid having a single [^{13}C] is 20 × 1.1% (22%). The odds of containing four are 22% × 22% × 22% × 22%, or approximately 0.2%, a figure usually considered to be negligible (Fig. 3.2). Further additions yield no practical advantage, and although the alteration in physical characteristics by the addition of each neutron is small, it is cumulative, so larger numbers have a greater impact on the physical and chemical characteristics of a compound with no added benefits.

Eicosanoids labeled with [^{2}H] make excellent internal standards, and many are available commercially (Cayman Chemical Co., Ann Arbor, MI; BIOMOL, Plymouth Meeting, PA). However, the label must be incorporated when the compound is synthesized, so generation of a new internal standard can be a challenging synthetic project. A relatively simple alternative is to incorporate [^{18}O] into the carboxylic acid group. This method involves formation of the methyl ester and subsequent cleavage by either Li^{18}OH (Murphy and Clay, 1982) or by an esterase (Pickett and Murphy, 1981) in the presence of H$_2$^{18}O (Cambridge Isotope Laboratories, Inc., Andover, MA). We have successfully applied the latter method to a wide range of eicosanoids with good success. The following observations apply:

1. The esterification/hydrolysis cycle may require several iterations in order to get a high degree of isotope labeling.
2. When the enzyme preparation is dried, it becomes hygroscopic; it should be exposed to air for the minimum amount of time possible.

3. An inexpensive enzyme is porcine liver esterase (Sigma). We have used several forms of this product with no noticeable difference in reactivity toward the eicosanoid esters.

4. Carboxyl oxygens are exchangeable, so internal standards labeled in this manner should not be exposed to esterases or extremes of pH. The naturally occurring esterases of plasma should be denatured before using this type of internal standard. Lowering the sample pH to the 3 to 4 range in order to protonate the acid and improve extraction efficiency does not cause loss of label.

5. The availability of $H_2^{18}O$ has been sporadic. There are often several degrees of purity available, and they vary little, if any, in cost.

When a stable isotope-labeled IS is used, several factors contribute to the accuracy and precision of the method:

1. Accuracy of IS concentration. It is difficult to quantitate a commercially obtained IS, so it is imperative that they be purchased from a reliable source. An IS prepared in-house should be compared by HPLC/MS/MS with a trusted unlabeled product. When a standard solution needs to be replaced, this solution and the replacement solution should each be compared to an unlabeled analogue in order to ensure that in-lab quantitation is temporally consistent.

2. New standards should be diluted to a practical concentration. Solutions that are too concentrated are susceptible to inaccuracies due to evaporation during use, and samples must be spiked with small volumes that are difficult to measure accurately and precisely. A 50-μg vial of standard can be accurately diluted to 50 ml, giving a concentration of 1 ng/μl. This is suitable for preparing a combination spike, or may be further diluted for a single spike.

3. Aliquoting the sample. Accurate and precise aliquots of urine are, with modern pipettes, to a large degree dependent on the technique of the individual.

4. Aliquoting the IS. This is the single most important step of the sample workup. The common air-displacement pipettes are not suitable for use with organic solvents. A positive displacement pipette (e.g., Microman, Rainen Instrument, LLC) is needed. The IS should be of such concentration that approximately 50 μl is added per milliliter of sample. Smaller volumes are not as accurate, and larger volumes may interfere with sample extraction. Mixing all the ISs for a set of samples is advisable, so that a single addition will carry all the spikes. After spiking, a sample should be mixed and allowed to equilibrate for at least 15 min. Small samples should be diluted to 1 ml with water in order to keep the organic component to 5%. Standards should be stored in organic solution at $-20°$ or below; acetonitrile is a good choice for almost all eicosanoids.

All stable isotope–labeled internal standards will contain a small amount of unlabeled material—the "blank." The blank level is often low enough to be negligible (less than 0.5%); however, if the unlabeled content is significant, it must be subtracted from the IS/analyte ratio measured by the MS.

3.2. Derivatization

Chemical alteration of eicosanoids is not necessary for ES MS analysis; however, some require modification for efficient reverse-phase chromatography. For the analytes discussed here, conversion of the ketone or aldehyde group to the methoxime (MO) derivative is sufficient to prevent any unwanted tautomerization (see Fig. 3.1A). Each MO adds 29 amu to the molecular weight of the compound and yields two stereoisomers, which may or may not be separated on the HPLC (Fig. 3.3). A convenient protocol involves addition of the reagent to the spiked sample before extraction. The MO concentration must be high in order for the reaction to proceed quickly. We add one-half the sample volume of a 1 g/ml aqueous solution of MO HCl (25 g into 25 ml; refrigeration is not necessary) and let stand for 15 min at room temperature. The pH of the resulting solution is low, ~3, and it can be directly applied to a solid phase extraction (SPE) cartridge. For mouse urine samples, which are usually approximately 0.1 ml, it is best to form the derivative and then dilute to 1 ml with water before SPE.

3.3. Extraction

Solid phase extraction has to a large degree replaced other methods of removing eicosanoids and their metabolites from aqueous matrices. Reasons for this include ease of use and smaller volumes of organic solvent use. This choice is optimal for HPLC/MS/MS analysis because a simple protocol can remove water, salt, protein, and organic contaminants that differ significantly in polarity from the target compounds from a urine sample, yielding a small volume of volatile organic solvent that, when evaporated, is suitable for direct HPLC/MS/MS injection. Analyte recovery is often greater than 90%. Specifically designed vacuum manifolds holding 12 or 24 cartridges are available (Supelco, Phenomenex). We have focused our efforts on

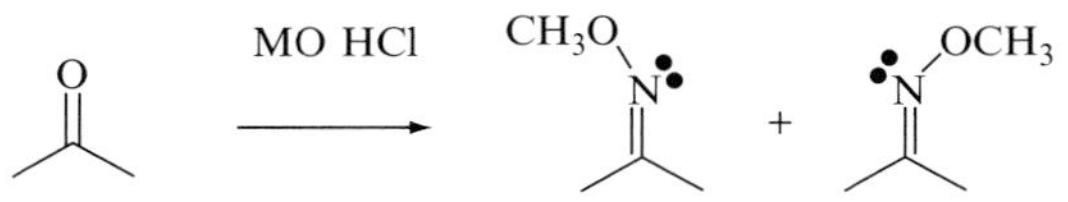

Figure 3.3 Conversion of a ketone or aldehyde to the MO derivative forms two stereo-isomers because of the unshared pair of electrons on the N.

developing a single, simple SPE method that works for a wide range of eicosanoids and metabolites that will allow future additions of target compounds and support our Lipidomics program. This protocol works equally well for urine samples of 1 or 2 ml.

Reverse-phase (C18) SPE cartridges have been widely used for eicosanoid extraction from aqueous medium, but recent advances in the technology have introduced polymeric sorbents with multifunctional coatings. Manufacturers claim better recovery and reproducibility, but their main advantages in routine preparation of large numbers of samples are their resistance to deconditioning and their ability to be dried completely. Thorough drying is important because if vestiges of water remain, it will elute with the sample and must then be either dried or the organic layer must be transferred to another vial. Either option adds time and complexity to the protocol. Polymeric sorbents can be dried completely by the application of house vacuum to the manifold for 15 min. Our lab uses StrataX reverse-phase cartridges (30-mg sorbent, 1-ml tube; Phenomenex). The protocol for extraction of urinary eicosanoid metabolites follows:

1. Condition cartridge with 1 ml of acetonitrile.
2. Equilibrate cartridge with 1 ml of water.
3. Apply sample.
4. Wash with 1 ml of 5% acetonitrile in water.
5. Dry with vacuum for 15 min.
6. Elute with 1 ml of 5% acetonitrile in ethyl acetate.

The eluate is dried under a gentle stream of nitrogen and dissolved in 10 μl of acetonitrile. Analytes may degrade in aqueous solution, so water is not added until the samples are to be injected. Then, 190 μl of HPLC grade water is added. The last step is filtration of the sample to remove any particulates, such as fines from the SPE cartridge, which may plug the HPLC lines. Small centrifugal filters with a 0.2-μm nylon membrane do an effective job (Costar Spin-X HPLC). One half of the sample is injected, with the remainder reserved in case reinjection should be required.

3.4. Chromatography

High-quality HPLC equipment is available from a number of manufacturers, most of which will have models capable of meeting the requirements of this assay; instrument choice may depend more on subjective factors such as personal experience or MS manufacturer. Reliability is the single most important consideration. Our lab has used Shimadzu equipment for many years, and we have found it to have a fine balance of dependability and affordability (LC-20AD, Shimadzu).

Autosamplers are obviously crucial for high-throughput operation, and there are several general types available. Designs that draw the sample directly into the injection syringe instead of into tubing connected to a remote syringe are less complex and easier to troubleshoot, maintain, and repair. Sample refrigeration is important for overnight operation. Autosamplers can be considered as stand-alone units since they may be controlled by contact closure signals, and therefore do not have to be purchased as part of an HPLC system.

Most manufacturers now offer vacuum degassing units that remove dissolved gasses from the mobile phase before it enters the pumps. This option is more effective and more convenient than helium sparging, and is highly advised for unattended operation.

Chromatographic columns are divided into two main modes of action. Normal-phase chromatography separates compounds based on interactions between the polar stationary phase and the polar functions of the analyte. Gradients increase the polarity of the mobile phase, so compounds elute in order of increasing polarity. Reverse-phase (RP) chromatography uses hydrophobic phases, the most common being C18, and gradients run from water to organic solvents, eluting compounds in order of increasing hydrophobicity. Although the urinary metabolites are relatively polar lipids due to chain shortening and oxidation, they keep enough of their lipid nature to be separated under RP conditions. Also, by buffering the mobile phase at pH 5.7, they can be chromatographed as the carboxylate anion, ready for MS detection.

The selection of an HPLC column is the most daunting task facing the analyst; a supplier's catalog may contain close to 100 phase choices. The chromatographer must make decisions concerning the mode (normal or reverse phase), length and diameter of the column, and the size of the particles, as well as mobile phase composition, pH, and gradient. These choices become more difficult as the number of analytes increases, because each analyte has its own chromatographic characteristics and array of common impurities, often including isomers, from which it must be separated. These factors will be considered in detail as each metabolite is discussed below. C18 phases from different suppliers vary dramatically in their ability to separate the various analytes. We have identified a stationary phase that is able to achieve all the chromatographic goals required by this set of metabolites (Luna C18(2) 3 μ, Phenomenex). All data shown here were obtained from this phase.

Decreasing the size of particles with which an HPLC column is packed improves the resolution (decreases the height equivalent to a theoretical plate [HETP]), but also increases the back pressure. We have found three-micron particles to be a workable compromise. Some manufacturers are introducing particles in the two-micron range, a move that may improve the resolution of eicosanoid assays or decrease the analysis time, but many older systems are not suited for the pressures involved. New acquisitions of HPLC equipment should depend on its ability to work reliably at the higher pressures required by these smaller particles. At this point in time, the selection of phases available in this size is limited.

Decreasing column diameter reduces the mobile phase flow rate through a column, so eluting peaks are narrower, sensitivity is higher, and solvent use is decreased. The limit here is imposed by the difficulty of packing small bore columns—generally those less than 2 mm ID—and their subsequent increased purchase cost. Systems must be plumbed with very low dead volume in order to take advantage of low flow rates and minimize gradient lag. Columns of 2 mm ID are the current compromise. They can be operated at 0.2 ml/min flow rate, which tolerates the dead volumes achieved within a well-plumbed system. PEEK tubing of 0.005″ ID with finger-tight fittings can be used throughout the system.

We have found that a column length of 150 mm is necessary to achieve the required separations for the metabolites discussed here. A column should stand up to several hundred urine extracts, keeping the cost per sample low. A pre-column filter will dramatically increase column life and decrease down-time due to plugged columns. A reasonable choice is one that minimizes dead volume by fitting directly into the column inlet, with easily replaced filter elements (e.g., SecurityGuard, Phenomenex).

Because of the complex nature of urine extracts, column performance will be preserved if each analysis ends with a column wash of at least 1 min with 100% solvent B. A solvent divert valve should redirect the column effluent to waste during the wash. An equilibration period of at least 15 min is required to get reproducible chromatography of the early-eluting peaks. A simple A/B column switching valve and a spare pump delivering the initial mobile phase can be set up so that the column equilibration is done off-line while a second column is in use. Switching is done through contact closure signals from the data system.

3.5. Mass spectrometry

We have found tandem quadrupole API instrumentation to be reliable, sensitive, and robust. Advances in electronics have produced instruments that, when situated in a temperature-controlled environment, need infrequent calibration once per month or even less frequently. Modern API interfaces are refractive to deposits of nonvolatile sample components, and if carefully prepared and filtered, hundreds of samples can be injected before any maintenance is required. The current generation of instruments is the first to be able to perform at this level, as demonstrated by the ability to quantitate 2,3-dinor-6-keto-PGF$_{1\alpha}$ reliably in human urine, usually the lowest concentration of this set of metabolites and the last to be transferred from E(/NI/GC/MS to HPLC/MS/MS. We would not be so presumptuous as to state that we have identified the best instrument available, but will simply state that our current instruments are able to achieve the goal of essentially around-the-clock sample analysis over long periods of time (Quantum, Thermo-Finnigan).

Electrospray of the anionic free acids was chosen over EC APCI, which, although it may offer more sensitivity, also adds to the complexity of the sample preparation protocol by requiring the formation of the PFB ester and the subsequent drying step. Our experience has shown that, for reasons that are not yet clear, the ability to form ions in EC APCI mode is highly instrument-specific; instruments from some manufacturers do not perform well in this mode (unpublished observations).

The optimum operational mode for the MS/MS is selected reaction monitoring (SRM). This method takes full advantage of the tandem quadrupole arrangement. The first quadrupole passes only the carboxylate anion (precursor ion) of the target compound forward into the collision chamber where it is fragmented by collision with gas atoms, usually argon. The product ions resulting from this collision are then guided to the second quadrupole, which filters a preselected ion. Each precursor/product pair is called a transition. Although the signal intensity of the product ion is less than that of the precursor, the signal-to-noise ratio, which is the ultimate determinant of sensitivity, is greatly increased. Clearly, the choice of product ions and collision energy is crucial, with the final choice depending on results from real samples. Collision energy is a "soft" parameter, and is not directly transferable even between instruments of the same model; it must be considered as a starting point for the tuning process. Instrument parameters used in our lab are listed below. Final tuning of the instrument should be done by infusing an eicosanoid under conditions approximating those of the final analysis, that is, the HPLC effluent should be at the correct flow rate and its makeup should be close to that of the gradient when analyte elution occurs. The position of the interface should be optimized in all three axes, and since HPLC flow rate, API probe temperature, desolvating gas flow and other parameters, depending on instrument model, are interactive, a thorough initial tune is a complex undertaking and should be performed with patience. Source collision-induced dissociation (SCID) on the quantum system critical for sensitivity and optimizes at approximately 12 V.

The MS parameters file is subdivided to minimize the number of transitions monitored at any one time. Each additional transition shortens the integration time, decreases the number of ions counted, and adversely affects the signal-to-noise ratio of the peaks.

3.6. Instrument operational parameters

We list here the general parameters involved in the analysis of a sample. Analyte-specific parameters are presented below.

3.6.1. HPLC

Column: Luna C18(2) 3 μ 150 × 2.00 mm (Phenomenex).
Prefilter: Security guard. Cartridges C18 4 × 2.0 mm (Phenomenex).
Mobile phase: Solvent A—water. Solvent B—95% acetonitrile/5% methanol. (A and B contain 0.005% acetic acid adjusted to pH 5.7 with NH_4OH.)
Flow rate: 0.200 ml/min.
Gradient: Injection at 5% B (hold 1 min); 11% B at 1.1 min; 17% B at 30 min; 21% B at 50 min; 30% B at 50.1 min; 40% B at 65 min (hold 5 min).
Wash: 100% B at 70.1 min (hold 3 min).
Equilibration: 5% B at 73.1 min (hold 15 min).
Oven temperature: 40°.

3.6.2. API

Spray voltage: 2000 V.
H–ESI vaporizer temperature: 240°.
Sheath gas pressure: 70 (arbitrary units).
Auxiliary gas flow: 5 (arbitrary units).

3.6.3. MS

Capillary temperature: 350°.
Source CID: 12V.
Ion sweep gas pressure: 0 (arbitrary units).

4. URINARY MARKERS OF SYSTEMIC EICOSANOID SYNTHESIS

Analytical parameters specific to each compound are noted.

4.1. PGE$_2$

Tetranor–PGEM is the major urinary metabolite of PGE_1 and PGE_2 in humans (Hamberg *et al.*, 1972); we also monitor this compound in mouse urine (Fig. 3.4). Synthetic tetranor–PGEM and a 17, 17', 18, 18', 19, 19'-d_6 internal standard are commercially available (Cayman). We spike with 25 ng/ml human urine and 10 ng/0.1 ml mouse urine. The MO derivative is required for acceptable chromatographic characteristics. Since this compound has two ketones, four MO isomers are created. Under our conditions, two major peaks are observed. Mouse urine samples

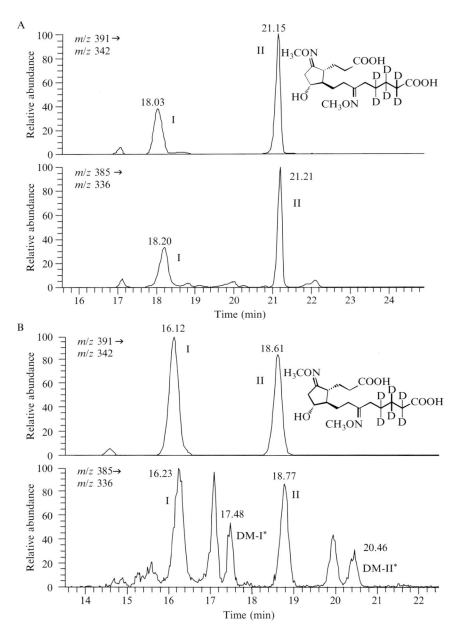

Figure 3.4 (A) Tetranor-PGEM in human urine. Upper panel: d_6-tetranor-PGEM MO. The four methoxime isomers present as two major peaks (I;II). Lower panel: Endogenous tetranor-PGEM MO. (B) Tetranor-PGEM in mouse urine. Upper panel: d_6-tetranor-PGEM MO. Lower panel: Endogenous tetranor-PGEM MO. The numerous seemingly isomeric peaks require rigorous chromatographic conditions for resolution. Some of the peaks ★ seem to originate from PGD_2 (see text).

contain additional HPLC peaks, and require an extremely shallow gradient for sufficient resolution to enable peak area determination. The relatively polar nature of tetranor-PGE M MO causes its interaction with reverse phase sorbents to be delicate, requiring thorough equilibration before sample injection; at least 15 min are recommended. Transitions monitored (12 to 30 min) are 385 → 336 for the endogenous compound and 391 → 342 for the internal standard. The collision energy is 15 eV.

4.2. PGD₂ (human)

In human urine we measure 2,3-dinor-11β-PGF$_{2\alpha}$ (Sweetman and Roberts, 1985). Authentic 2,3-dinor-11β-PGF$_{2\alpha}$ is available commercially (Cayman), but there is no stable isotope-labeled internal standard available; [^{18}O$_2$] 2,3-dinor-11β-PGF$_{2\alpha}$ is produced using the method above from unlabeled standard. We spike with 2 ng/ml urine. Transitions monitored are 325 → 145 for the endogenous compound and 329 → 145 for the internal standard (30 to 57 min). The collision energy is 13 eV. There are numerous compounds in urine that meet the criteria of this transition and stringent chromatographic conditions are required to get suitably resolved peaks (Fig. 3.5).

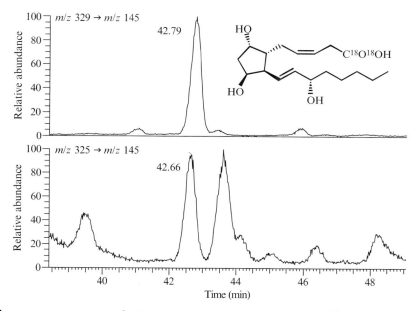

Figure 3.5 2,3-dinor-11β-PGF$_{2\alpha}$ in human urine. Upper panel: [^{18}O$_2$] 2,3-dinor-11 β-PGF$_{2\alpha}$. Lower panel: Endogenous 2,3-dinor-11β-PGF$_{2\alpha}$ (42.66 min). All samples contain several nearby peaks that are separated only by long retention times.

4.3. PGD₂ (mouse)

Very little is known about the metabolism of PGD$_2$ in the mouse. We observe two peaks in the tetranor-PGEM region of the mass chromatogram* that behave as isomers and seem to originate from PGD$_2$ (unpublished observations). We postulate that they are the analogous tetranor metabolites with the D-ring instead of the E-ring. We are pursuing this assumption, and investigating their use as an index of PGD$_2$ production (see Fig. 3.4). This compound is not available commercially. All HPLC/MS/MS parameters are identical to those of tetranor-PGEM.

4.4. TxA₂ (mouse)

The mouse TxA$_2$ metabolite monitored is 2,3-dinor-TxB$_2$. There is no stable isotope-labeled internal standard available commercially, so the unlabeled product (Cayman) is converted to the [$^{18}O_2$] form using the technique described above. We spike with 1 ng/0.1 ml urine. 2,3-dinor-TxB$_2$ exists in equilibrium between an open form and the closed ring (hemiacetal) form, necessitating formation of the MO derivative for acceptable chromatographic characteristics (Fig. 3.6). The transitions monitored are $370 \rightarrow 155$ for the

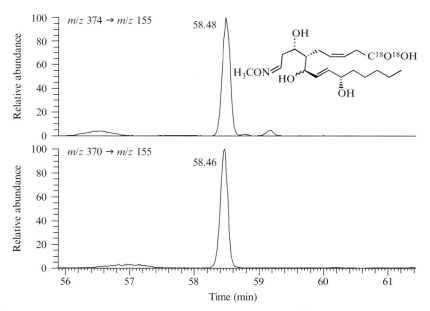

Figure 3.6   2,3-dinor-TxB$_2$ in mouse urine. Upper panel: [$^{18}O_2$] 2,3-dinor-TxB$_2$ MO. Lower panel: Endogenous 2,3-dinor-TxB$_2$ MO.

endogenous compound and 374 → 155 for the internal standard (30 to 60 min). The collision energy is 15 eV.

4.5. TxA$_2$ (human)

The major urinary metabolite of TxA$_2$ in humans is 11-dehydro-TxB$_2$ (Catella and FitzGerald, 1987). Authentic 11-dehydro-TxB$_2$ and a 3, 3', 4, 4'-d$_4$ internal standard are available commercially (Cayman). We spike with 5 ng/ml urine. Allowing the sample to equilibrate at pH ∼3 for 1 h after addition of MO HCl drives the compound toward the lactone form (Schweer *et al.*, 1987). Studies evaluating the time course of this reaction are ongoing. Transitions monitored are 367 → 305 for the endogenous compound and 371 → 309 for the internal standard (57 to 70 min) at collision energy 16 eV (Fig. 3.7).

4.6. PGI$_2$

The urinary metabolite 2,3-dinor-6-keto-PGF$_{1\alpha}$ is the metabolite of choice to monitor PGI$_2$ in human urine (Falardeau *et al.*, 1981). We also monitor it in mouse urine (Fig. 3.8), where we observe an earlier-eluting compound that shares at least three product ions. We are investigating the possibility

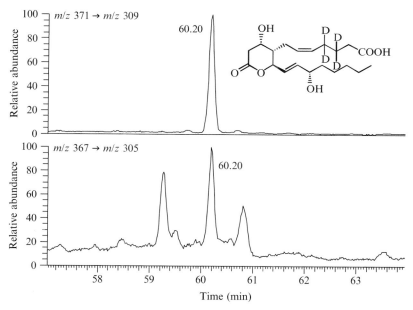

Figure 3.7 11-dehydro-TxB$_2$ in human urine. Upper panel: d$_4$–11-dehydro-TxB$_2$. Lower panel: Endogenous 11-dehydro-TxB$_2$.

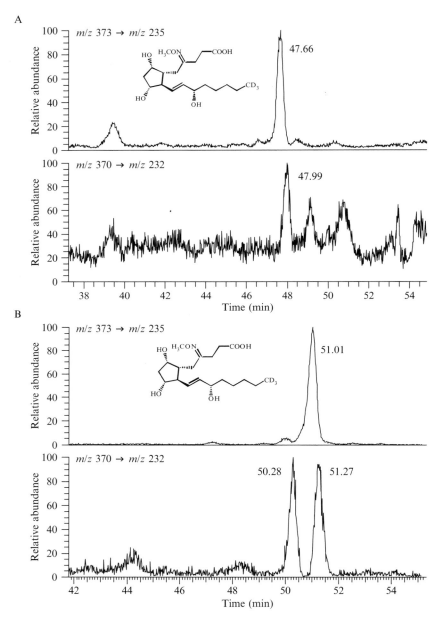

Figure 3.8 (A) 2,3-dinor-6-keto-PGF$_{1\alpha}$ in human urine. Upper panel: d$_3$-2,3-dinor-6-keto-PGF$_{1\alpha}$ MO. Lower panel: Endogenous 2,3-dinor-6-keto-PGF$_{1\alpha}$ MO. (B) 2,3-dinor-6-keto-PGF$_{1\alpha}$ in mouse urine. Upper panel: d$_3$-2,3-dinor-6-keto-PGF$_{1\alpha}$ MO. Lower panel: Endogenous 2,3-dinor-6-keto-PGF$_{1\alpha}$ MO. The earlier eluting peak seems to be an isomer (see text).

that this is an enzymatically derived isomer of authentic 2,3-dinor-6-keto-$PGF_{1\alpha}$. At pH 5.7, this compound exists in equilibrium with the hemiketal form, requiring formation of the MO derivative for acceptable chromatography (see Fig. 3.1A). Authentic 2,3-dinor-6-keto-$PGF_{1\alpha}$ and a 20, 20, 20-d_3 internal standard are available (BIOMOL). We spike with 2 ng/ml human urine and 5 ng/0.1 ml mouse urine. Transitions monitored are $370 \rightarrow 232$ for the endogenous compound and $373 \rightarrow 235$ for the internal standard (30 to 57 min). The collision energy is 20 eV.

4.7. Isoprostanes

Isoprostanes (iPs) are nonenzymatic products arising from the autooxidation of arachidonic acid (Liu *et al.*, 1999). They have been shown to correlate with syndromes of oxidant stress in cardiovascular pathology (Lawson *et al.*, 1999). We have published an HPLC/MS/MS assay for the major urinary iP in humans, 8, 12-*iso*-$iPF_{2\alpha}$-VI (Li *et al.*, 1999), and it is incorporated into this assay (Fig. 3.9). We have synthesized a tetradeuterated internal standard (Lawson *et al.*, 1999). Although this d_4 compound is not available commercially, d_{11}-8,12-*iso*-$iPF_{2\alpha}$-VI is available (Cayman). We spike with 10 ng/ml human urine, and 1 ng/0.1 ml mouse urine. Transitions monitored are $353 \rightarrow 115$ for the endogenous compound and $357 \rightarrow 115$ for the internal standard (57 to 70 min). The collision energy is 24 eV.

5. DATA ANALYSIS AND INTERPRETATION

Although all LC/MS/MS data systems include software for finding and integrating peaks, they are often unable to determine correct peak start/end times when the signal-to-noise ratio is low and/or incompletely resolved peaks are present. Subjective judgment must play a part, and all traces integrated by the data system must be inspected and corrected, if warranted.

Urinary levels of eicosanoid metabolites should be normalized to creatinine levels in order to adjust for urine production.

6. CONCLUDING REMARKS

We have described in detail a method for the quantitation of urinary metabolites of eicosanoids that are of particular interest in cardiovascular research. It allows the measurement of metabolites of PGE_2, PGD_2, PGI_2, TxB_2, and a representative iP in a 0.1-ml mouse urine sample or in 1 ml of human urine after the addition of a spike containing the five stable

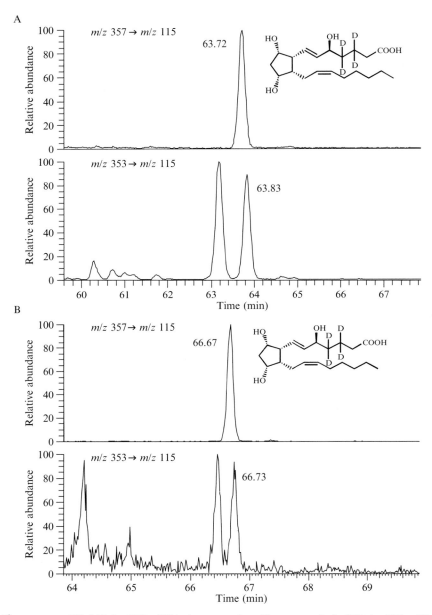

Figure 3.9 (A) 8,12-*iso*-iPF$_{2\alpha}$-VI in human urine. Upper panel: d$_4$-8,12-*iso*-iPF$_{2\alpha}$-VI. Lower panel: Endogenous 8,12-*iso*-iPF$_{2\alpha}$-VI. (B) 8,12-*iso*-iPF$_{2\alpha}$-VI$_\alpha$ in mouse urine. Upper panel: d$_4$-8,12-*iso*-iPF$_{2\alpha}$-VI. Lower panel: Endogenous 8,12-*iso*-iPF$_{2\alpha}$-VI.

isotope-labeled internal standards, derivatization *in situ*, and extraction on an SPE cartridge. The wide range of polarity of the metabolites and a multitude of endogenous compounds, often isomeric, requires a long analysis time (70 min), but this must be compared to previous GC/MS methods, which required five separate samples, stringent two-plate TLC purifications for each, two additional derivatization steps per sample, and five GC/MS analyses.

Our future plans are to add compounds to the list of analytes as needs warrant and conditions allow. We will continue to evaluate HPLC phases and particle sizes in an effort to shorten the instrument cycle time while preserving the necessary separations. We expect that future versions of the assay will be able to monitor more compounds in less time and work with smaller volumes of urine as tandem MS instrumentation increases in sensitivity and HPLC technology advances.

REFERENCES

Catella, F., and FitzGerald, G. A. (1987). Paired analysis of urinary thromboxane B_2 metabolites in humans. *Thromb. Res.* **47,** 647–656.

Falardeau, P., Oates, J. A., and Brash, A. R. (1981). Quantitative analysis of two dinor urinary metabolites of prostaglandin I2. *Anal.Biochem.* **115,** 359–367.

Greindling, K. K., and FitzGerald, G. A. (2003). Oxidative stress and cardiovascular injury. Part I: Basic mechanisms and *in vivo* monitoring of ROS. *Circulation* **108,** 1912–1916.

Grosser, T., Fries, S., and FitzGerald, G. A. (2006). Biological basis for the cardiovascular consequences of COX-2 inhibition: Therapeutic challenges and opportunities. *J. Clin. Invest.* **116,** 4–15.

Hamberg, M. (1972). Inhibition of prostaglandin synthesis in man. *Biochem. Biophys. Res. Commun.* **49,** 720–726.

Hunt, D. F., Stafford, G. C., Jr., Crow, F. W., and Russell, J. W. (1976). Pulsed nositive negative ion chemical ionization mass spectrometry. *Anal. Chem.* **48,** 2098–2105.

Lawson, J. A., Li, H., Rokach, J., Adiyaman, M., Hwang, S. W., Khanapure, S. P., and FitzGerald, G. A. (1998). Identification of two major F2 isoprostanes, 8,12-iso- and 5-epi-8, 12-iso- isoprostane F2alpha-VI, inhuman urine. *J. Biol. Chem.* **273,** 29295–29301.

Lawson, J. A., Rokach, J., and FitzGerald, G. A. (1999). Isoprostanes: Formation, analysis and use as indices of lipid peroxidation *in vivo. J. Biol. Chem.* **274,** 24441–24444.

Lee, S. H., Williams, M. V., DuBois, R. N., and Blair, I. A. (2003). Targeted lipidomics using electron capture atmospheric pressure chemical ionization mass spectrometry. *Rapid Commun. Mass Spectrom.* **17,** 2168–2176.

Li, H., Lawson, J. A., Reilly, M., Adiyaman, M., Hwang, S. W., Rokach, J., and FitzGerald, G. A. (1999). Quantitative high performance liquid chromatography/tandem mass spectrometric analysis of the four classes of F_2-isoprostanes in human urine. *Proc. Natl. Acad. Sci. USA* **96,** 13381–13386.

Liu, T., Stern, A., Roberts, L. J., and Morrow, J. D. (1999). The isoprostanes: Novel prostaglandin-like products of the free radical-catalyzed peroxidation of arachidonic acid. *J. Biomed. Sci.* **6,** 226–235.

Min, B. H., Pao, J., Garland, W. A., de Silva, J. A. F., and Parsonnet, M. (1980). Determination of an antisecretary trimethyl prostaglandin E_2 analog in human plasma by combined capillary column gas chromatography-negative chemical ionisation mass spectrometry. *J. Chromatogr. Biomed. Appl.* **183,** 411–419.

Morrow, J. D., Awad, J. A., Boss, H. J., Blair, I. A., and Roberts, L. J., 2nd (1992). Non-cyclooxygenase-derived prostanoids (F2-isoprostanes) are formed *in situ* on phospholipids. *Proc. Natl. Acad. Sci. USA* **89,** 10721–10725.

Murphy, R. C., and Clay, K. L. (1982). Preparation of ^{18}O derivatives of eicosanoids for GC-MS quantitative analysis. *Methods Enzymol.* **86,** 547–551.

Patrono, C., Rodriguez, L. A. Garcia, and Baigent, C. (2005). Low-dose aspirin for the prevention of atherothrombosis. *N. Engl. J. Med.* **353,** 2373–2383.

Pickett, W. C., and Murphy, R. C. (1981). Enzymatic preparation of carboxyl oxygen-18 labeled prostaglandin $F_{2\alpha}$ and utility for quantitative mass spectrometry. *Anal. Biochem.* **111,** 115–121.

Reilly, M., and FitzGerald, G. A. (2002). Gathering intelligence on antiplatelet drugs: The view from 30 000 feet. *BMJ* **324,** 59–60.

Schweer, H, Meese, C. O., Furst, O, Kuhl, P. G., and Seyberth, H. W. (1987). Tandem mass spectrometric determination of 11-dehydrothromboxane B_2, an index metabolite of thromboxane B_2 in plasma and urine. *Anal. Biochem.* **164,** 156–163.

Singh, G., Gutierrez, A., Xu, K., and Blair, I. A. (2000). Liquid chromatography/electron capture atmospheric pressure chemical ionization/mass spectrometry: Analysis of pentafluorobenzyl derivatives of biomolecules and drugs in the attomole range. *Anal. Chem.* **72,** 3007–3013.

Sweetman, B. J., and Roberts, L. J., 2nd (1985). Metabolic fate of endogenously synthesized prostaglandin D_2 in a human female with mastocytosis. *Prostaglandins* **30,** 383–400.

CHAPTER FOUR

Lipidomics in Diabetes and the Metabolic Syndrome

Richard W. Gross*,† *and* Xianlin Han*

Contents

Abstract

Shotgun lipidomics, based on multi-dimensional mass spectrometric array analyses after multiplexed sample preparation and intrasource separation, has been recently advanced to a mature technique for the rapid and reproducible global analysis of cellular lipids. At its current stage, this technology enables us to analyze more than 20 lipid classes and thousands of individual lipid molecular species directly from lipid extracts of biologic samples. Following a brief introduction to the foundations underlying this rapidly expanding technology, we present detailed protocols used for the identification and quantitation of plasma triglycerides, determination of the human heart lipidome, and analysis of cellular cardiolipin molecular species. Through the use and practice of shotgun lipidomics, new insights into the cardiovascular pathobiology manifest in diabetes and the metabolic syndrome can be accrued.

* Division of Bioorganic Chemistry and Molecular Pharmacology, and Department of Internal Medicine, Washington University School of Medicine, St. Louis, Missouri
† Department of Molecular Biology and Pharmacology, Washington University School of Medicine, and Department of Chemistry, Washington University, St. Louis, Missouri

Methods in Enzymology, Volume 433
ISSN 0076-6879, DOI: 10.1016/S0076-6879(07)33004-8

1. INTRODUCTION

The major diseases of the early 21st century in industrialized nations are largely related to caloric excess, high-fat diet, and sedentary lifestyle. This unfortunate combination results in a plethora of metabolic abnormalities that collectively predispose to atherosclerosis, myocardial infarction, congestive heart failure, stroke, and hypertension (Miranda *et al.*, 2005; Moller and Kaufman, 2005; Unger, 2002). The synergistic deleterious effects of pathologic metabolic states such as obesity, diabetes and hyperlipidemia on atherosclerosis, inflammation, and hypertension have recently been linked and collectively termed the metabolic syndrome (Miranda *et al.*, 2005; Moller and Kaufman, 2005; Unger, 2002). In early work, chronic elevations of serum lipids were shown to ultimately result in changes in cellular lipid metabolism that led to intracellular lipid accumulation and resultant compromise of organ function. Increasing focus has been placed on measurement of the elevation of triglycerides and other lipids not only in serum, but also in specific tissues susceptible to accumulation of lipids that compromise physiologic function. This has necessitated the development of robust technologies that can identify and accurately measure hundreds of lipids and characterize alterations in the kinetics of their metabolism during the onset and progression of clinical disease as well as their response to treatment. While identification of the major molecular species of cellular lipids by ion beam mass spectrometry began over two decades ago, pioneering advances in ionization technology made by Dr. Fenn have now been refined to dramatically improve the accuracy, dynamic range, and sensitivity of mass spectrometry to facilitate the study of the alterations in lipid metabolism that underlie the metabolic syndrome in patients. Through the use of chemical strategies that allow identification and quantitation of hundreds of critical lipids, these advances are beginning to impact on our understanding of diabetes, diabetic cardiomyopathy, and the downstream pathologic sequelae of the metabolic syndrome. The goal of this chapter is to first introduce the concept of multidimensional mass spectrometry in shotgun lipidomics and next to provide a detailed description of its use and practice in understanding diabetes and the metabolic syndrome.

2. BRIEF INTRODUCTION OF MULTI-DIMENSIONAL MS-BASED SHOTGUN LIPIDOMICS

One of the major new developments in current lipidomics is multidimensional mass spectrometry (MS)–based shotgun lipidomics (Han and Gross, 2001, 2005b; Han *et al.*, 2004b). This approach has now evolved into a mature technology that includes a series of simple steps such as multiplexed extractions for sample preparation, intrasource separation to resolve lipid classes based on their electrical propensities, multidimensional MS, and array analyses. Application of this process allows the facile identification and quantitation of individual lipid molecular species through multiple ratiometric comparisons of both internal and external molecular species.

The lipids of each biological sample (commonly containing 50 to 500 μg of protein mass content from cell, tissue, or biologic fluid) can be extracted by solvent(s) under acidic, basic, and/or neutral conditions (i.e., multiplexed extractions). The ESI ion source behaves like an electrophoretic cell and can selectively separate different charged moieties under high electrical potential (typically ~4 kV; Gaskell, 1997; Ikonomou et al., 1991). Since lipid classes possess different electrical properties, largely depending on the nature of their polar head groups (Han and Gross, 2003, 2005a), the electrospray ion source can be used to resolve lipid classes in a crude lipid extract based on the intrinsic electrical properties of each lipid class (now termed "intrasource separation of lipids"; Han and Gross, 2005a,b; Han et al., 2004b). A successful strategy for the intrasource separation of lipid classes from crude lipid extracts has been previously discussed in detail (Han and Gross, 2003, 2005a,b; Han et al., 2004b). Through this approach, a comprehensive series of mass spectra can readily be obtained from the lipid extract of a biological sample (e.g., human heart; Fig. 4.1).

After intrasource separation, each ion peak in the mass spectrum of interest represents at least one (and very often two or more) lipid molecular species, particularly those acquired by mass spectrometers having only low mass-resolution capabilities. We recognized that most of these biological lipid species are linear combinations of aliphatic chains, lipid backbones (e.g., glycerol, sphinganine, and sphingosine), and/or head groups, each of which represents a building block of the lipid molecular species under consideration. For example, three moieties linked to the hydroxyl groups of glycerol can be recognized as three individual building blocks and if each building block is identified, then each individual glycerol-derived lipid molecular species in a given sample can be determined (Han and Gross, 2005b). An analogous approach can also be used to define other lipid classes (e.g., sphingomyelins in which the phosphocholine head group, the sphingoids [long chain bases], and the fatty acyl amides represent the three building blocks of each molecular species) (Han, 2007; Han and Gross, 2005b). Identification of these building blocks can be accomplished by two powerful tandem MS techniques (i.e., neutral loss (NL) scanning and precursor ion (PI) scanning) that monitor the specific loss of a neutral fragment or the yield of a fragment ion, respectively, each of which represents a specific building block. Therefore, all the building blocks of

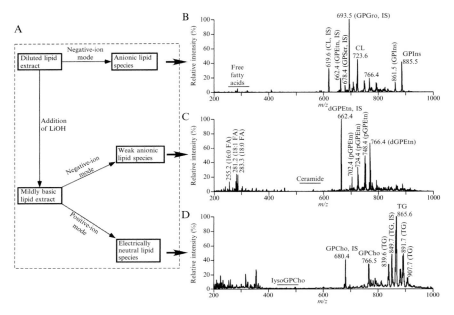

Figure 4.1 ESI/MS analysis of a lipid extract of human heart after intrasource separation. Panel A illustrates a schematic diagram of the procedures used with intrasource separation, that is, to analyze lipids in the positive- or negative-ion mode in the absence or presence of a small amount of LiOH. Panels B through D show the corresponding mass spectra acquired under these three conditions, each of which displays a distinct lipid profile of different lipid classes. Other abbreviations for lipids are provided in the text. IS, internal standard.

each lipid class constitute additional dimensions to the molecular ions present in the original mass spectrum, which is referred to as the first dimension. By determining the building blocks present in a given primary molecular ion in the first dimension with those in the second dimension (obtained by NL and PI scanning), the chemical structure(s) (including regiospecificity) and the presence of isobaric constituents in the original molecular ion can be determined (Han and Gross, 2005b). Specific example applications are given below.

After identification of each individual molecular ion of a lipid class, quantitation by shotgun lipidomics is performed using a two-step ratiometric procedure (Han and Gross, 2005b; Han *et al.*, 2004a). First, the abundant and nonoverlapping molecular species of a class are quantified by comparison with a preselected internal standard of the class after ^{13}C de-isotoping (Han and Gross, 2001, 2005a). Next, some or all of the determined molecular species of the class (plus the preselected internal standard) are used as standards to determine the mass content of other low-abundance or overlapping molecular species using one or multiple MS/MS traces (each of which represents a specific building block of the class of interest) by two-dimensional (2D) MS. Through this second step of quantitation, the linear dynamic range can be dramatically extended by eliminating background noise and by filtering the overlapping molecular species through a multidimensional process (Han and Gross, 2005a).

Through lipid class–selective intrasource ionization and subsequent 2D MS analyses, shotgun lipidomics, at its current stage, enables us to fingerprint and quantify individual molecular species of most major and many minor lipid classes in cellular lipidomes, which collectively represent more than 95% of the total lipid mass (composed of as many as 1000 molecular species), directly from their CHCl₃ extracts. These classes of lipids include choline glycerophospholipid (GPCho), ethanolamine glycerophospholipid (GPEtn), phosphatidylinositol (GPIns), phosphatidylglycerol (GPGro), phosphatidylserine (GPSer), phosphatidic acid (GPA), sphingomyelin (SM), galactosylceramide, glucosylceramide, sulfatide, free fatty acid (FFA), triacylglycerol (TG), lysoGPCho, lysoGPEtn, lysoGPA, acylcarnitine, cholesterol and cholesterol esters, and ceramide (Cer) (including dihydroceramide). Cardiolipin (CL) (Han *et al.*, 2006), sphingosine-1-phosphate, and dihydrosphingosine-1-phosphate (Jiang and Han, 2006) are the newest lipid classes that have been added to the list of lipids identified and quantified by shotgun lipidomics.

3. Profiling Human Plasma Triacylglycerol Molecular Species

Human plasma contains many different TG molecular species that represent the metabolic signatures of lipid metabolism integrated through multiple organs that are regulated by the nutritional history and metabolic demands of the organism. Therefore, the TG molecular species present in plasma and/or fractionated lipoproteins from plasma, may be used as potential biomarkers for the presence metabolic syndrome and for the identification of biochemical mechanisms responsible for its pathogenesis. Profiling of these TG molecular species can be readily achieved by our shotgun lipidomics technology as outlined above and previously described in detail (Han and Gross, 2001).

The building blocks of TG molecular species are the three acyl chains of which there are approximately 10 prominent species present in nature. Characterization of lithiated TG molecular species demonstrates that these acyl chains can be defined by neutral loss of the fatty acids corresponding to the acyl chains in the various TG molecular species (Han *et al.*, 2000; Hsu and Turk, 1999). Therefore, neutral loss scanning of all naturally occurring fatty acids followed by array analysis of the cross-peaks in a 2D-MS process facilitates identification of the component of each molecular ion. In this 2D mass spectrum, the first dimension (i.e., the survey scan) is represented by a specific molecular ion of the lithium adducts of TG molecular species present in the lipid extract, while the second dimension displays the individual aliphatic chain building blocks present in all detectable TG species (i.e., all of the masses of neutral loss fragments; Han and Gross, 2001). The cross-peaks of a given lithiated TG molecular ion in the first dimension with the building blocks present in the second dimension represent the fatty acyl chains in the TG molecule. The presence of multiple isobaric molecular species of TG in each molecular ion can thus be readily determined from the number and intensities of these underlying neutral loss fragment peaks in combination with the mass of the parent TG molecular ion (Han and Gross, 2001).

Human blood samples obtained from volunteers following an approved study protocol were collected into a cryovial (6 ml) in the presence of dextrose-sodium citrate buffer solution. Each blood sample was centrifuged at 800 rpm for 10 min at room temperature. The upper clear, yellow layer was transferred to a 15-ml conical plastic centrifuge tube by using a plastic transfer pipette. This upper layer was further centrifuged at 2300 rpm at room temperature for 25 min, and the resulting supernatant was collected as the plasma sample.

A human plasma sample (100 μl, ~7 mg of protein, which is determined through a protein assay for each individual sample) was transferred into a borosilicate glass culture tube (size 16×100 mm). Lipids from each plasma sample were extracted by a modified Bligh and Dyer procedure (Bligh and Dyer, 1959) in the presence of internal standards as described previously (Cheng $et\ al.$, 2006). Specifically, to each plasma sample, 4 ml of $CHCl_3$/MeOH (1:1, v/v), 1.7 ml of LiCl solution (50 mM), and internal standard (2 nmol T17:1 TG/mg of protein) were added. The extraction mixtures were vortexed and centrifuged at 2500 rpm for 5 min. The $CHCl_3$ layer of each extract mixture was carefully removed and saved. An additional 2 ml of $CHCl_3$ was added to the MeOH/aqueous layer of each test tube. After centrifugation, the $CHCl_3$ layer from each individual sample was combined and dried under a nitrogen stream. Each individual lipid residue was then resuspended in 4 ml of $CHCl_3$/MeOH (1:1), back-extracted against 1.8 ml of 10-mM aqueous LiCl, and the resulting $CHCl_3$ extract was dried as described above. Each individual residue was then resuspended in ~1 ml of $CHCl_3$ and filtered with a 0.2-μm polytetrafluoroethylene syringe filter and dried under a nitrogen stream. Finally, each individual residue was resuspended in $CHCl_3$/MeOH (1:1, v/v) corresponding to 0.1 ml/mg of protein. The lipid extracts were finally flushed with nitrogen, capped, and stored at $-20°$ for ESI/MS analyses (typically conducted within 1 week).

The lipid extracts were further diluted greater than 50-fold with 1:1 $CHCl_3$/MeOH prior to direct injection into the ESI ion source of a TSQ Quantum Ultra mass spectrometer (ThermoElectron, San Jose, CA).

The dilution procedure is to ensure that the lipid concentration is less than 100 pmol/μl in 1:1 CHCl$_3$/MeOH to avoid lipid aggregation during analysis (Han, 2007). In addition, through extensive extraction against aqueous LiCl, filtering with a small-pore-size filter, and a final dilution procedure, contaminants that might interfere with ionization stability and efficiency are substantially reduced for lipid analysis.

Next, the diluted lipid extract with addition of a small amount of LiOH/ MeOH solution was directly infused into the ESI source of the mass spectrometer at a flow rate of 4 μl/min with a syringe pump. Typically, a 2-min period of signal averaging in the profile mode was employed for each survey scan in the positive-ion mode in the mass range for TG analysis (i.e., m/z 750 to 960). A customized sequence subroutine operated under Xcalibur software was employed for automatic acquisition of neutral loss scans (2 min each) corresponding to all potentially naturally occurring fatty acids. In the positive-ion neutral-loss scanning mode, the first and third quadrupoles serve as independent mass analyzers using a mass resolution setting of peak width 0.5 Th, while the second quadrupole serves as a collision cell. A collision gas pressure of 1.0 mTorr and a collision energy of 35 eV were employed. These conditions were determined through product–ion analysis to achieve fragment peak intensities resulting from the neutral loss of a fatty acid from the sn-2 position of a TG species essentially equals those resulting from the neutral losses of fatty acids from both sn-1 and sn-3 positions of the TG species within experimental error.

Thus, the total ion currents (TIC) of these neutral loss scans of the entire study constitutes a TIC chromatogram (Fig. 4.2A), which directly reflects the mass composition of individual fatty acyl chains in the TG pool of interest. Averaging of all acquired mass spectra in the segment corresponding to each neutral loss followed by smoothing of the averaged trace by employing the built-in program of the Xcalibur software gives rise to a final neutral loss scan. Fig. 4.2B shows a representative scan resulting from neutral loss of 18:3 FA. All these neutral loss scans together with the survey scan constitute a 2D mass spectrum (Fig. 4.3). Analysis of the cross-peaks between a lithiated TG molecular species in the survey scan and its corresponding building blocks in the second dimension allowed us to identify individual TG molecular species including those that were isomeric. For example, the primary ion at m/z 837.7 is composed of 14:0, 16:0, 16:1, 18:0, 18:1, and 18:2 FA building blocks (indicated by the broken line at the left in Fig. 4.3). Since a lithiated TG molecular species at m/z 837.7 must contain 53 total carbons with 2 double bonds or 54 total carbons with 9 double bonds, and the ion intensities resulting from the neutral loss of the three acyl chains from a given TG molecule are nearly equal, isomeric TG molecular species of 14:0–18:1–18:1 TG, 14:0–18:0–18:2 TG, 16:0–16:0–18:2 TG, and 16:0–16:1–18:1 TG can be identified. Other TG molecular species corresponding to other primary ions were also similarly identified.

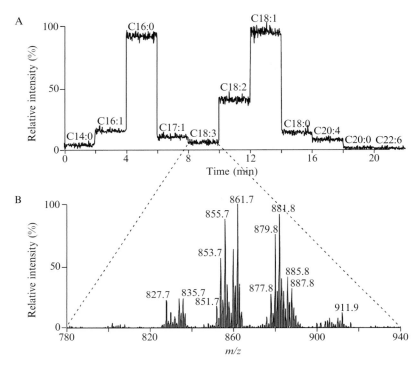

Figure 4.2 Total ion current chromatogram of stepwise scanning of naturally occurring fatty acids neutrally lost from TG molecular species in a human plasma lipid extract and an example of an averaged mass NL spectrum of 278 u (i.e., 18:3 FA). The lipid extract from a human plasma sample (100 μl) was prepared by a modified Bligh and Dyer procedure as described in the text and were analyzed in the positive-ion mode after infusion of the diluted lipid extract in the presence of a small amount of LiOH at a flow rate of 4 μl/min. The stepwise NL scanning was performed by a sequential and customized program operating under Xcalibur software. Each segment of individual NL scanning was taken for 2 min in the profile mode (panel A). Panel B shows an example of the NL mass spectrum averaged from all of scans acquired in the segment corresponding to NL of 278 u (i.e., 18:3 FA). For tandem mass spectrometry in the NL mode, both the first and third quadrupoles were coordinately scanned with a mass difference (i.e., NL) corresponding to the NL of a nonesterified fatty acid from TG molecular species, while collisional activation was performed in the second quadrupole. NL, neutral loss; u, mass unit.

Once individual TG molecular species were identified, quantitation of these TG molecular species was performed as follows. The ion current of each identified TG species was obtained by averaging the total ion currents of three fragments resulting from the neutral loss of the three fatty acids from the lithiated TG species. This ion current was corrected for an ionization efficiency factor that is predetermined based on the number of carbons present and the degree of unsaturation of the molecular species as described

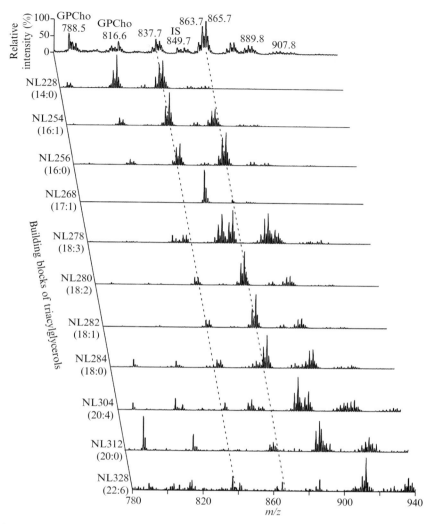

Figure 4.3 Two-dimensional mass spectrometric analyses of triacylglycerol molecular species in a lipid extract of a human plasma sample. The preparation of the lipid extract from a human plasma sample and the generation of each NL mass spectrum are described in the legend of Fig. 4.2. NL scanning of all naturally occurring aliphatic chains (i.e., the building blocks of TG molecular species) of a human plasma CHCl₃ extract were utilized to determine the identities of each molecular ion, deconvolute isobaric molecular species, and quantify individual TG molecular species by comparisons with a selected internal standard (i.e., T17:1 TG, shown in the NL scanning trace of NL268.2). Each MS/MS trace of the two-dimensional ESI mass spectra was acquired by sequentially programmed custom scans operating under Xcalibur software as shown in Fig. 4.2. All displayed mass spectral traces are normalized to the base peak in each trace. NL, neutral loss.

previously (Han and Gross, 2001) prior to de-isotoping of ^{13}C isotopologues. It should be pointed out that the isotopologues from other atoms such as ^{18}O, ^{15}N, and ^{2}H can be ignored in comparison to the inherent experimental errors and biologic variability, but can be similarly considered if one would like to include these effects as well. The corrected and ^{13}C de-isotoped ion current was then compared with that of the selected internal standard to obtain the mass content of this TG species.

4. ANALYSIS OF HUMAN HEART LIPIDOME

Human heart samples were obtained from patients undergoing cardiac transplantation for chronic nonischemic congestive heart failure. Human myocardial wafers were pulverized into a fine powder at the temperature of liquid nitrogen. A myocardial sample (~10 mg) was weighed from each heart sample and homogenized in 1 ml of ice-cold phosphate-buffered saline (diluted 10 times with water) with a Potter-Elvehjem tissue grinder. Protein assays on each individual homogenate were performed. Each myocardial sample (~1 ml of homogenate) was transferred to a disposable borosilicate glass tube (16 × 100 mm) and a premixed internal standard solution including 12, 14, 3, 7.5, 0.03, 1.5, 2, 2, 1.5, 1.5, 0.75, 1.5, and 4 nmol/mg of protein of 14:1–14:1 GPCho, 15:0–15:0 GPEtn, 15:0–15:0 GPGro, T17:1 TG, N17:0 ceramide, d_4–16:0 FA, N12:0 SM, 14:0–14:0 GPSer, 14:0 lysoGPEtn, 17:0 lysoGPCho, 14:0–14:0 GPA, 17:1 monoacyl-glycerol, and T14:0 CL, respectively, were added based on protein concentration. These internal standards allowed us to normalize the final quantified lipid content to the protein mass present and to eliminate the effects of potential losses from incomplete sample recovery during processing, and were selected because they represent less than 0.1% of the endogenous cellular lipid mass as demonstrated by ESI/MS lipid analysis. Lipids from each homogenate were prepared by using a modified Bligh and Dyer procedure as described in the last section. Each lipid extract was reconstituted with a volume of 500 μl/mg of protein (which was based on the original protein content of the samples as determined from protein assays) in $CHCl_3$/MeOH (1:1, v/v). The lipid extracts were finally flushed with nitrogen, capped, and stored at −20° for ESI/MS analyses (typically within 1 week).

By using intrasource separation, many of the lipid classes in each human heart lipid extract were typically profiled under three ionization conditions in the absence or presence of a low concentration of LiOH (see Fig. 4.1). We first acquired a survey scan (Panel B) in the negative-ion mode at a scan rate of 1.5 s per scan after direct infusion of a 50-fold diluted plasma lipid extract at a flow rate of 4 μl/min for 1 min. Then, by using either neutral loss (NL) or precursor-ion (PI) scanning for 2 min after collision-induced dissociation, each of the building blocks corresponding to the individual molecular species of an anionic lipid class (or a group of anionic lipid classes) of interest in the survey scan region was determined separately. This procedure was repeated for the analysis of all anionic lipid classes of interest in the selected mass regions. In fact, both the survey scan and the CID scans of all building blocks of all anionic lipid classes were automatically acquired by using an Xcalibur customized sequence subroutine. Each scan was averaged from all acquired spectra in a corresponding segment of the sequence similar to that shown in Fig. 4.2A.

Next, by arrangement of the building blocks based on fragment size, a 2D mass spectrum was constructed. For example, a 2D mass spectrum for identification of individual GPIns molecular species in human plasma lipid extracts was constructed as shown in Fig. 4.4. The common molecular species of the GPIns class are in the m/z range of 800 to 950, and its building blocks include fatty acyl carboxylates and head groups (i.e., inositol phosphate, m/z 241, and glycerophosphate derivative, m/z 153), both of which can be analyzed by using precursor-ion scanning. Analysis of the cross-peaks of a parent ion in the survey scan with its corresponding building blocks in the second dimension allowed us to identify individual GPIns molecular species including isomeric species and their regiospecificity. For example, the primary ion at m/z 861.7 is crossed with the building blocks of 18:2 FA (PI279), 18:0 FA (PI283), inositol phosphate (PI241), and glycerophosphate derivative (PI153) (see the broken line in Fig. 4.4). Because the cross-ion peak intensity of 18:2 FA was more abundant than that of 18:0 FA, this ion was identified as deprotonated 18:0–18:2 GPIns. Similarly, the ion at 885.7 was identified as 18:0–20:4 GPIns. These identification processes were automatically performed by using a program based on macros functions of the Excel software which will be described elsewhere in detail. These identified GPIns molecular species were quantified by using a two-step procedure. Briefly, molecular species of 18:0–18:2 and 18:0–20:4 GPIns at m/z 861.7 and 885.7 were quantified by direct comparison to the selected internal standard (i.e., 15:0–15:0 GPGro) after ^{13}C de–isotoping. The mass levels of other low abundance GPIns molecular species including 16:0–18:2, 16:0–18:1, 18:0–18:1, 18:1–20:4, 18:0–22:4, and 18:0–22:5 GPIns at m/z 833.7, 835.7, 859.7, 883.7, 909.7, and 911.7, respectively, were determined by using a PI241 scan and employing both 18:0–18:2 and 18:0–20:4 GPIns as standards. Again, this quantitation process was automated by our Excel macros–based program. It should be pointed out that the selected internal standard is likely not optimal for the analysis of GPIns molecular species. If a proper GPIns analogue is made available, this currently used standard will be replaced. However, the ionization efficiency of GPGro and GPIns are not significantly different as previously demonstrated (Han and Gross, 1994).

Individual molecular species of other anionic lipid classes (as well as chlorinated GPCho) were all identified by a procedure similar to that used

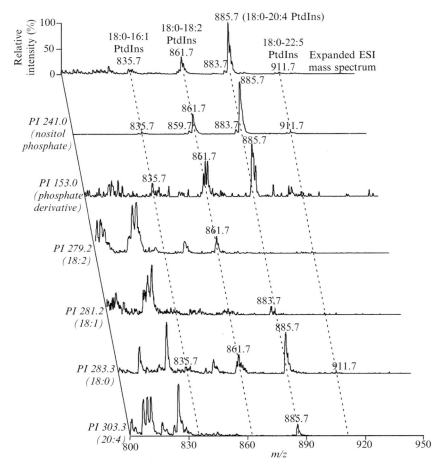

Figure 4.4 Two-dimensional mass spectrometric analysis of GPIns molecular species in a mouse myocardial lipid extract. Lipids were extracted from mouse myocardium by a modified Bligh and Dyer procedure as previously described (Han *et al.*, 2005b). Each MS or MS/MS trace of the 2D ESI mass spectrum was acquired by sequentially programmed, customized scans operating under Xcalibur software. For negative-ion tandem mass spectrometry in the precursor-ion (PI) mode, the first quadrupole was scanned in the selected mass range, and the second quadruple was used as a collision cell, while the third quadrupole was fixed to monitor the ion of interest (i.e., inositol phosphate, glycerophosphate, or a fatty acyl carboxylate fragmented from molecular species of GPIns). All mass spectral traces were displayed after being normalized to the base peak in each individual trace. IS, internal standard.

for GPIns with an identical customized sequence. Only the head group building blocks (i.e., PI241 and PI153) in the 2D mass spectrum for identification of GPIns was replaced by NL87 (neutral loss of serine) and PI153 for GPSer molecular species, PI153 (glycerophosphate) for GPGro,

GPA, and lysoGPA molecular species, and NL50 (neutral loss of chloromethane) for chloride adducts of GPCho under the experimental conditions. Anionic lipid molecular species quantitation was similarly performed to that for GPIns molecular species using a two-step procedure as described above. It should be emphasized that PI153 is not specific to the stated lipid classes, but is present in all anionic phospholipid classes. However, the peak intensities of the ions corresponding to GPGro and GPA in the PI153 scan were considerably more sensitive in comparison to those of GPIns and GPSer in addition to the differences in the m/z range. PI153 from CL molecular species may overlap with GPGro and GPA, but its double-charged character (discussed in detail in the next section), can assist in distinguishing these anionic lipid classes. It should be pointed out that NL50 for GPCho molecular species may be replaced by neutral loss of methyl acetate (or formate) if the lipids were extracted against an acetate- (or formate-) containing medium.

After the analysis of anionic lipids, a small amount of LiOH in MeOH was added to the diluted lipid solution. The amount of LiOH should be determined based on the mass levels of GPEtn or GPCho plus TG depending on which one is higher since OH^- is used to deprotonate the GPEtn molecular species in the negative-ion mode and Li^+ is used as an adduct for the analysis of GPCho and TG molecular species in the positive-ion mode. The mass levels of GPCho and GPEtn can be estimated from the protein mass content of a sample of interest. For example, the mass contents of GPCho or GPEtn are typically at the level of 100-nmol lipid/mg protein in the heart samples.

Next, two survey scans in the m/z range of 200 to 1000 were acquired in the negative- and positive-ion modes after direct infusion of the diluted lipid solution with addition of LiOH at a flow rate of 4 μl/min (see Fig. 4.1C and D). Lithium adducts of GPCho molecular species were identified through 2D mass spectrometry utilizing NL183 (phosphocholine) and the neutral loss scans of all potential naturally occurring fatty acids. This analysis was redundant since GPCho molecular species were previously analyzed as chloride adducts in the negative-ion mode as described above. LysoGPCho molecular species were directly identified from either NL183 as lithium adducts or NL59 (trimethylamine) as sodium adducts since the aliphatic moiety could be determined from the m/z value. SM molecular species were also directly identified from NL183, assuming the sphingoid base is sphingosine, which accounts for the majority of the mass content of total sphingoid base. If identification of lower abundance sphingoid bases is a priority, neutral loss of various neutral fragments can be performed as previously described (Hsu and Turk, 2005). TG molecular species in human heart lipid samples were similarly analyzed as described in the previous section.

Although the abundant molecular species of GPEtn were readily visualized in the survey scan acquired in the negative-ion mode (see Fig. 4.1C), their identities could be confirmed by using 2D mass spectrometric analysis in which the building blocks include all of the fatty acyl carboxylates resulting from GPEtn molecular species. However, the low abundance GPEtn molecular species are overlapped by other molecular species of anionic lipid classes and the fragment corresponding to the GPEtn head group is not sufficiently sensitive to determine these low abundance molecular species. Therefore, we utilized Fmoc derivatization of the ethanolamine head group to identify these molecular species as described previously (Han *et al.*, 2005a). Briefly, equal molar amounts of Fmoc chloride and DMAP were added to the diluted lipid solution immediately prior to injection into the mass spectrometer. The neutral loss scan of Fmoc (i.e., NL222) shows all the amine-containing molecular species including lysoGPEtn. Identification of these Fmoc-derivatized GPEtn molecular species was achieved by analyzing their building blocks including lysoplasmenylethanolamine fragments through a 2D process as described (Han *et al.*, 2005a). Again, quantitation of the identified molecular species of GPCho, SM, lysoGPCho, GPEtn, and lysoGPEtn was similarly performed to that of GPIns molecular species by using a two-step procedure as described above.

Three features of the lipid composition of end-stage diseased human hearts seem evident. First, the majority of ethanolamine glycerophospholipids present in myocardium are plasmalogen molecular species. A substantive portion (20%) of the choline glycerophospholipids are also plasmalogens. Second, arachidonic acid is the major *sn*-2 aliphatic constituent in plasmalogen molecular species in human myocardium, thus rendering these species suitable substrates for phospholipases which generate eicosanoid second messengers during signal transduction. Third, end-stage human myocardium contains abundant amounts of triglycerides. These TG molecular species are predominantly composed of unsaturated fatty acids and include the presence of odd–chain–length fatty acids that presumably are synthesized by peroxisomal α oxidation (Su *et al.*, 2004).

5. HIGH MASS RESOLUTION–BASED SHOTGUN LIPIDOMICS FOR THE ANALYSIS OF CARDIOLIPIN MOLECULAR SPECIES

There is no specific fragment which represents the head group of cardiolipin molecular species following fragmentation by CID. However, we exploited the double-charged characteristic of cardiolipin molecular species which possess two phosphodiesters per cardiolipin molecule. Fortunately, other double-charged lipid ions are rarely found in the m/z range for cardiolipin. Moreover, the neutral loss of ketenes from double-charged cardiolipin molecular ions to yield double-charged triacyl monolysocardiolipins is specific to cardiolipin molecular species. Therefore, we have recently developed shotgun lipidomics techniques for the identification and quantitation of cardiolipin molecular species using this unique chemical property (Han *et al.*, 2006).

Through the double-charged nature of cardiolipins, ions from each cardiolipin individual molecular species (including the overlapping and low-abundance molecular species) can be directly recognized through searching for the presence of the $[M-2H+1]^{2-}$ isotopologue peaks by using a mass spectrometer with a resolving power high enough to detect the double-charged ions. For example, Fig. 4.5 shows the mass spectrum of a mouse-heart lipid extract acquired in the negative-ion mode by using a TSQ Quantum Ultra Plus mass spectrometer (ThermoElectron, San Jose, CA) with a mass resolution setting of 0.3 Th. The double-charged cardiolipin molecular ion peaks are well displayed (see inset of Fig. 4.5).

After identification of these unique cardiolipin molecular ion peaks, product-ion analyses of the plus-one isotopologue peaks can be performed

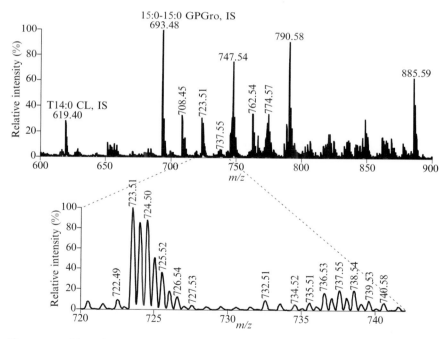

Figure 4.5 Negative-ion ESI mass spectrometric analysis of double-charged cardiolipin molecular species in a lipid extract of mouse myocardium by a QqQ-type mass spectrometer with a high mass-resolution setting of FHMW 0.3 Th. CL, double-charged cardiolipin; GPGro, phosphatidylglycerol; IS, internal standard. (From Han, X., Yang, K., Yang, J., Cheng, H., and Gross, R. W. (2006). Shotgun lipidomics of cardiolipin molecular species in lipid extracts of biological samples. *J. Lipid Res.* **47**, 864–879, with permission of American Society for Biochemistry and Molecular Biology, Inc., © 2006.)

to confirm the identities of acyl chains of each individual cardiolipin molecular species. However, 2D MS analyses of all cardiolipin molecular species can be readily conducted by neutral loss analysis of the ketenes from double-charged cardiolipin molecular species to yield double-charged monolysocardiolipin (i.e., the specific building blocks of cardiolipin) in the m/z region of interest.

Once the cardiolipin molecular ion is recognized by searching for the $[M-2H+1]^{2-}$ isotopologue peak of double-charged ions and identified through 2D MS analyses, quantitation of cardiolipin molecular species can be readily performed by ratiometric comparison of the de-isotoped ion peak intensities of the cardiolipin molecular species with a selected cardiolipin internal standard. Importantly, we have demonstrated a 1000-fold dynamic range and an accurate linear correlation between the cardiolipin molecular species (Han et al., 2006). The ^{13}C de-isotoping based on the $[M-2H+1]^{2-}$ isotopologue peak intensity can be performed using the following equation:

$$I_{total} = I_1 \times (92.42/n + 1 + 5.41 \times 10^{-3}(n-1)$$
$$+ 1.95 \times 10^{-5}(n-1)(n-2)$$
$$+ 5.3 \times 10^{-8}(n-1)(n-2)(n-3) + \ldots)$$

where I_{total} is the de-isotoped ion intensity of an individual cardiolipin molecular species (M) of interest; I_1 is the peak intensity of its $[M-2H+1]^{2-}$ isotopologue; and n is the total carbon numbers in the species. In the case of the presence of a cardiolipin molecular species (M-2) that has two mass units less than M, I_1 should represent the corrected ion peak intensity of the plus–one isotopologue of M. The correction can be calculated as follows:

$$I_{1(M)} = I'_{1(M)} - I_{3(M-2)} = I'_{1(M)} - 1.95 \times 10^{-5}I_{1(M-2)}(n-1)(n-2)$$

where $I'_{1(M)}$ is the determined peak intensity at the $[M-2H+1]^{2-}$ isotopologue of M; $I_{3(M-2)}$ is the peak intensity of the plus–three isotopologue (i.e., $[(M-2)-2H+3]^{2-}$) of M-2 cardiolipin molecular species; and $I_{1(M-2)}$ is the peak intensity of the plus–one isotopologue of M-2 cardiolipin molecular species. By applying this newly developed technique, accurate quantitation of an unprecedented number of cardiolipin molecular species was accomplished from a mouse myocardial lipid extract (Han et al., 2006).

ACKNOWLEDGMENT

Supported by National Institutes of Health grant P01 HL57278.

REFERENCES

Bligh, E. G., and Dyer, W. J. (1959). A rapid method of total lipid extraction and purification. *Can. J. Biochem. Physiol.* **37,** 911–917.

Cheng, H., Guan, S., and Han, X. (2006). Abundance of triacylglycerols in ganglia and their depletion in diabetic mice: Implications for the role of altered triacylglycerols in diabetic neuropathy. *J. Neurochem.* **97,** 1288–1300.

Gaskell, S. J. (1997). Electrospray: Principles and practice. *J. Mass Spectrom.* **32,** 677–688.

Han, X. (2007). Neurolipidomics: Challenges and developments. *Front. Biosci.* **12,** 2601–2615.

Han, X., and Gross, R. W. (1994). Electrospray ionization mass spectroscopic analysis of human erythrocyte plasma membrane phospholipids. *Proc. Natl. Acad. Sci. USA* **91,** 10635–10639.

Han, X., and Gross, R. W. (2001). Quantitative analysis and molecular species fingerprinting of triacylglyceride molecular species directly from lipid extracts of biological samples by electrospray ionization tandem mass spectrometry. *Anal. Biochem.* **295,** 88–100.

Han, X., and Gross, R. W. (2003). Global analyses of cellular lipidomes directly from crude extracts of biological samples by ESI mass spectrometry: A bridge to lipidomics. *J. Lipid Res.* **44,** 1071–1079.

Han, X., and Gross, R. W. (2005a). Shotgun lipidomics: Electrospray ionization mass spectrometric analysis and quantitation of the cellular lipidomes directly from crude extracts of biological samples. *Mass Spectrom. Rev.* **24,** 367–412.

Han, X., and Gross, R. W. (2005b). Shotgun lipidomics: Multi-dimensional mass spectrometric analysis of cellular lipidomes. *Expert Rev. Proteomics* **2,** 253–264.

Han, X., Abendschein, D. R., Kelley, J. G., and Gross, R. W. (2000). Diabetes-induced changes in specific lipid molecular species in rat myocardium. *Biochem. J.* **352,** 79–89.

Han, X., Cheng, H., Mancuso, D. J., and Gross, R. W. (2004a). Caloric restriction results in phospholipid depletion, membrane remodeling and triacylglycerol accumulation in murine myocardium. *Biochemistry* **43,** 15584–15594.

Han, X., Yang, J., Cheng, H., Ye, H., and Gross, R. W. (2004b). Towards fingerprinting cellular lipidomes directly from biological samples by two-dimensional electrospray ionization mass spectrometry. *Anal. Biochem.* **330,** 317–331.

Han, X., Yang, K., Cheng, H., Fikes, K. N., and Gross, R. W. (2005a). Shotgun lipidomics of phosphoethanolamine-containing lipids in biological samples after one-step *in situ* derivatization. *J. Lipid Res.* **46,** 1548–1560.

Han, X., Yang, K., Yang, J., Cheng, H., and Gross, R. W. (2006). Shotgun lipidomics of cardiolipin molecular species in lipid extracts of biological samples. *J. Lipid Res.* **47,** 864–879.

Hsu, F. F., and Turk, J. (1999). Structural characterization of triacylglycerols as lithiated adducts by electrospray ionization mass spectrometry using low-energy collisionally activated dissociation on a triple stage quadrupole instrument. *J. Am. Soc. Mass Spectrom.* **10,** 587–599.

Hsu, F. F., and Turk, J. (2005). Analysis of sphingomyelins. *In* "The Encyclopedia of Mass Spectrometry" (R. M. Caprioli, ed.), Vol. 3. Elsevier, New York.

Ikonomou, M. G., Blades, A. T., and Kebarle, P. (1991). Electrospray-ion spray: A comparison of mechanisms and performance. *Anal. Chem.* **63,** 1989–1998.

Jiang, X., and Han, X. (2006). Characterization and direct quantitation of sphingoid base-1-phosphates from lipid extracts: A shotgun lipidomics approach. *J. Lipid Res.* **47,** 1865–1873.

Miranda, P. J., Defronzo, R. A., Califf, R. M., and Guyton, J. R. (2005). Metabolic syndrome: Definition, pathophysiology, and mechanisms. *Am. Heart J.* **149,** 33–45.

Moller, D. E., and Kaufman, K. D. (2005). Metabolic syndrome: A clinical and molecular perspective. *Annu. Rev. Med.* **56,** 45–62.

Su, X., Han, X., Yang, J., Mancuso, D. J., Chen, J., Bickel, P. E., and Gross, R. W. (2004). Sequential ordered fatty acid a oxidation and D9 desaturation are major determinants of lipid storage and utilization in differentiating adipocytes. *Biochemistry* **43,** 5033–5044.

Unger, R. H. (2002). Lipotoxic diseases. *Annu. Rev. Med.* **53,** 319–336.

CHAPTER FIVE

LC-MS-MS Analysis of Neutral Eicosanoids

Philip J. Kingsley *and* Lawrence J. Marnett

Contents

Abstract

The neutral arachidonic acid derivatives N-arachidonoyl ethanolamine (anandamide or AEA), and 2-arachidonoylglycerol (2-AG) have been identified as endogenous ligands for the cannabinoid receptors. Additionally, these compounds have been identified as substrates of the second isoform of the cyclooxygenase enzyme (COX-2). Through the action of COX-2 and downstream prostaglandin synthases, a diverse family of prostaglandin glycerol esters (PG-Gs) and prostaglandin ethanolamides (PG-EAs) have been identified. Sensitive and reliable analytical methodology is crucial for the continued research on the biological roles of this family of lipids. In this chapter, we discuss methods for analyzing both the precursor endocannabinoids and their PG-like products by LC-MS-MS. Cation coordination provides the ionization, and selected reaction monitoring is successfully employed to provide a method of analysis that is both sensitive and specific.

Departments of Biochemistry, Chemistry, and Pharmacology, Vanderbilt Institute of Chemical Biology, Center in Molecular Toxicology, Vanderbilt-Ingram Cancer Center, and Vanderbilt University School of Medicine, Nashville, Tennessee

Methods in Enzymology, Volume 433
ISSN 0076-6879, DOI: 10.1016/S0076-6879(07)33005-X

1. INTRODUCTION

In the mid-1990s, two endogenous, neutral arachidonic acid (AA) derivatives that serve as cannabinoid receptor ligands were identified. These compounds, called endocannabinoids, are N–arachidonoylethanolamine (anandamide or AEA; Devane *et al.*, 1992) and 2-arachidonoylglycerol (2-AG; Mechoulam *et al.*, 1995; Sugiura *et al.*, 1995), and in each case, the carboxylic acid moiety of AA is replaced with a neutral functionality (Fig. 5.1). In the relatively short time since their discovery, endocannabinoids have been implicated in the regulation of a range of mammalian physiological processes, including feeding behavior (Di Marzo *et al.*, 2001), neural development (Berrendero *et al.*, 1999), retrograde neural signaling (Wilson and Nicoll, 2001), immune cell activation (Berdyshev, 2000), and emetic response (Van Sickle *et al.*, 2005). Additionally, it has been shown that 2-AG and AEA undergo bis-dioxygenation mediated by the second isoform of the cyclooxygenase enzyme, COX-2 (Kozak *et al.*, 2000; Yu *et al.*, 1997). Much like arachidonic acid is converted to

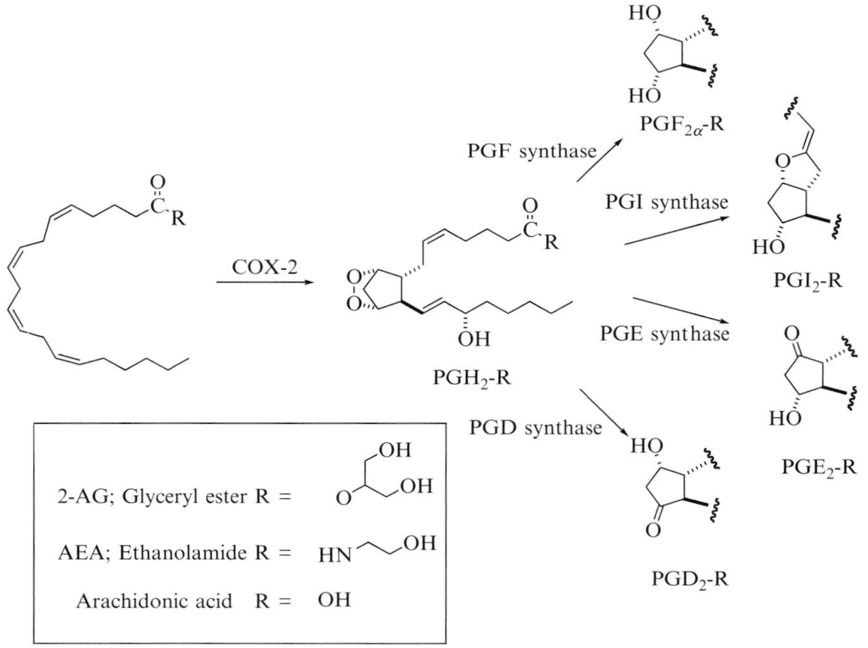

Figure 5.1 Structure of arachidonic acid and the endocannabinoids 2-arachidonoylglycerol (2-AG) and anandamide (AEA), and their conversion by COX-2 and various PG synthases to prostaglandin glyceryl esters and prostaglandin ethanolamides, respectively.

prostaglandin H_2 (PGH_2) and then transformed to various prostaglandins and thromboxane products through the action of specific synthases, COX-2 produces PGH_2-G and PGH_2-EA from 2-AG and AEA, respectively. The PGH_2 analogues are then substrates for most of the same synthases as PGH_2, resulting in a family of prostaglandin glycerol esters (PG-Gs) and prostaglandin ethanolamides (PG-EAs; see Fig. 5.1).

PG-Gs and PG-EAs have been shown to exhibit biological activity. PGE_2-G is produced in peritoneal macrophages harvested from CD-1 mice when the macrophages are treated with bacterial lipopolysaccharide (LPS) and zymosan. PGE_2-G is also produced in the macrophage cell line RAW264.7 following treatment with LPS and ionomycin (Rouzer and Marnett, 2005). $PGF_{2\alpha}$-EA, PGE_2-EA, and PGD_2-EA (PGE_2-EA and PGD_2-EA were quantitated as a single compound) have been observed after the administration of AEA in both FAAH $-/-$ and wild-type mice (Weber et al., 2004). PGE_2-G stimulates Ca^{2+} mobilization in RAW264.7 cells (Nirodi et al., 2004), which suggests that these novel COX-2 metabolites possess potentially important biological activities. Given the growing appreciation of the physiologic importance of endocannabinoids, and the exciting possibility that their oxygenated metabolites are biologically relevant, reliable extraction and quantitation techniques for these compounds are vital.

Several methods are reported in the literature for the purification and analysis of endocannabinoids from tissue. Most methods use either a Folch or Bligh and Dyer extraction to remove the analytes from the tissue of interest. From this point, derivatization is often employed, either to aid detection (by adding a flourophore) or to improve gas chromatography performance. Both Sugiura et al. (2001) and Kondo et al. (1998) describe derivatizing AEA and 2-AG with 1-anthroyl cyanide. The derivatives are then analyzed by high-performance liquid chromatography (HPLC) with fluorescence detection.

Berrendero et al. (1999) describe a trimethylsilyl derivatization of endocannabinoids, while Schmid et al. (2000) report a t-butyldimethylsilyl derivatization. In each case, the derivatives are analyzed via gas-chromatography, electron-ionization mass spectrometry (GC-EIMS), and selected ion monitoring (SIM) is performed on a derivative fragment (Berrendero et al., 1999; Schmid et al., 2000) or on the molecular ion (Berrendero et al., 1999). Analysis of the underivatized analytes is also reported. Maccarrone et al. (2001) describe analyzing AEA and 2-AG via GC-EIMS. SIM is performed on underivatized analyte fragments. Alternatively, Koga et al. (1997) use liquid-chromatography, atmospheric-pressure chemical ionization MS (LC-APCI-MS) analysis for the quantitation of AEA, observing the $[M+H]^+$ ion in SIM mode.

Our laboratory has developed a method for analyzing 2-AG and AEA, which employs silver cation coordination and liquid-chromatography, electrospray-ionization, tandem mass spectrometry (LC-ESI-MS-MS) detection (Kingsley and Marnett, 2003). The neutrality of endocannabinoids is a roadblock to their analysis by mass spectrometric methods. However, the 4 double bonds of the arachidonate backbone of these compounds make them rich in π electrons. Silver cations introduced into the mobile phase complex with these π electrons of 2-AG and AEA, forming an $[M+Ag]^+$ complex that is amenable to electrospray ionization and tandem mass spectrometric techniques.

Unlike endocannabinoids, there is a paucity of analytical methods for the analysis of PG-Gs and PG-EAs. Our laboratory has published a method for the simultaneous analysis of prostaglandin free acids and analogous PG-Gs (Kingsley et al., 2005), and Weber et al. (2004) have described the analysis of $PGF_{2\alpha}$-EA. Both methods describe HPLC separation of analytes and tandem mass spectrometric detection (in the form of selected reaction monitoring) in the positive ion mode. The methodology developed in our laboratory (Kingsley et al., 2005) involves complexing the neutral PG-Gs and PG-EAs with either the ammonium cation or a proton. The resultant $[M+NH_4]^+$ or $[M+H]^+$ complexes yield multiple intense fragments upon CID, several of which may be employed in selected–reaction monitoring (SRM). These techniques will be discussed in detail in this chapter. Together, they allow metabolic profiling of all the oxygenated metabolites derived from 2-AG and AEA by COX-2.

1.1. Materials and equipment

1.1.1. Materials

2-AG, AEA, 1(3)-PGE$_2$ glyceryl ester, 1(3)-PGD$_2$ glyceryl ester, 1(3)-PGF$_{2\alpha}$ glyceryl ester, PGE$_2$ ethanolamide, PGD$_2$ ethanolamide, PGF$_{2\alpha}$ ethanolamide, octadeuterated 2-AG and AEA, and tetradeuterated PGF$_{2\alpha}$ ethanolamide were purchased from Cayman Chemical (Ann Arbor, MI). Additionally, the glyceryl esters of PGE$_2$, PGD$_2$, PGF$_{2\alpha}$, the pentadeuterated analogues of PGE$_2$-G, PGD$_2$-G, PGF$_{2\alpha}$-G, and 6-keto-PGF$_{1\alpha}$-G, and tetradeuterated PGE$_2$ ethanolamide were synthesized as described previously (Kozak et al., 2000). All solvents used were HPLC grade.

1.1.2. Equipment

Mass spectral analyses were performed on either a ThermoFinnigan TSQ 7000 or a Quantum triple quadrupole instrument equipped with an electrospray source and operated in positive ion mode (ThermoFinnigan, San Jose, CA). LC-ESI-MS-MS analyses were performed with a Waters (Milford, MA) 2690 Separations Module operated in-line with the TSQ 7000 or a Surveyor pump and autosampler coupled to the Quantum. Liquid chromatographic separations were performed on reverse-phase narrow-bore columns that were typically held at 40°. Both gradient and isocratic elution were used for analysis. Specific details regarding mass spectrometer settings and chromatographic conditions will be given in the relevant sections.

Analytes were usually purified via solid-phase extraction, with Waters Sep-Pak silica cartridges (1 cc, 100 mg) used for endocannabinoid purification, and OASIS HLB cartridges (1 cc, 30 mg) used for the PG-like compounds.

2. PURIFICATION OF ANALYTES FROM CELL CULTURE MEDIUM AND TISSUE

2.1. Endocannabinoids

The tissue of interest is homogenized in a volume of organic solvent (40 ml/g of tissue); 9:1 ethyl acetate:hexane and 9:1 ethyl acetate:methanol have been used in our laboratory. Tissues homogenize with greater ease in the ethyl acetate:methanol solvent. Homogenization is carried out in a Tenbroeck tissue grinder at room temperature. The tissue to be analyzed and an aliquot of internal standard solution containing 2-AG-d_8 and/or AEA-d8 is added to the tissue grinder at the beginning of homogenization. A small electric motor can be used to assist homogenization. A metal rod is placed in the chuck of the motor and a size-2 rubber stopper is placed on the metal rod. This stopper fits tightly into the pestle of 40-ml and 15-ml Tenbroeck grinders, which allows the motor to rotate the pestle and thus hasten the homogenization of the sample. However, care must be taken to not let the motor rotate too rapidly to avoid heating the sample. If further processing of the sample will not occur immediately, the homogenate is stored at $-20°$.

After homogenization, the homogenate is centrifuged (it may also be washed with 30% its volume of water). The supernatant is transferred to a clean vessel and evaporated to dryness in a warm H_2O bath (temperature less than $40°$). The dried extracts are then purified via solid phase extraction (SPE) in a procedure similar to that described by Schmid et al. (2000). Briefly, the dried extracts are reconstituted in about 1 ml of chloroform and applied directly to unconditioned Si SPE cartridges. The cartridges are washed with 3×1.0 ml chloroform and eluted with 4×1.0 ml 2% methanol in chloroform. The eluent is evaporated to dryness. The samples are reconstituted in acetonitrile or methanol and injected onto the LC-MS-MS system. It has been noted in our lab that using a commercially available SPE manifold has resulted in sample contamination. Thus, we suspend the SPE cartridges in test tubes. Endocannabinoids may be similarly extracted from cell media. Ethyl acetate may be used to remove the analytes from the media by a liquid–liquid extraction. If further purification is desired, the sample may be dried, reconstituted in chloroform, and subjected to the described Si SPE.

While endocannabinoids are very strongly retained by reverse-phase (RP) solid-phase cartridges, it has been our experience that the method described by Schmid and modified as indicated above is superior. The main advantage is that Si SPE provides samples that exhibit a much lower baseline on the described LC-MS-MS system when compared to RP SPE methods. Also, solvent handling is quicker in the normal phase setting because of the high volatility of chloroform.

2.2. PG-Gs and PG-EAs

Solid-phase extraction via OASIS HLB (1 cc 30 mg cartridges) is used to purify analytes from cell medium. The medium to be analyzed is spiked with the appropriate deuterated internal standards and acidified, if so desired. Acidifying the medium allows the retention of prostaglandin free acids on the OASIS HLB cartridges. The medium is applied to an OASIS HLB cartridge which has been conditioned with 1 ml of methanol followed by 1 ml H_2O (or 0.5% aqueous acetic acid if the media was acidified). The loaded cartridge is then washed with 1 ml of 0.5% acetic acid (aq.) and 1 ml 0.5% acetic acid (aq.) with 15% methanol. Air is drawn through the cartridges for 1 to 2 min under reduced pressure and the analytes are eluted with 1 to 1.5 ml methanol. The eluent is then evaporated to dryness and reconstituted in 1:2 acetonitrile:water. The wash steps may be altered in response to the particular needs of the assay in use. We have found that the PG-like analytes start to elute from the OASIS HLB cartridges when the percentage of methanol in 0.5% acetic acid reaches 25%. In addition, the purification need not be accomplished on OASIS HLB cartridges. These compounds are well retained by reverse-phase modalities, and thus, the above method may be modified to accommodate many commercially available reverse-phase cartridges.

2.3. LC-MS analysis

2.3.1. Endocannabinoids

The ability of the silver cation to complex with double bonds (Winstein and Lucas, 1938) has long been known and exploited for the analysis of lipids. It has been used as a component of the stationary phase for silver chromatography, and has also been employed as a mode of ionizing compounds for mass spectral analysis (Canty and Colton, 1994; Havrilla et al., 2000; Morris, 1966; Nichols, 1952). The arachidonate backbone of 2-AG and AEA provides ideal binding sites for the silver cation, resulting in a $[M+Ag]^+$ entity that easily undergoes conversion to the gas phase via an ESI source.

When 2-AG is infused in 1:1 $MeOH:H_2O$ with 170 μM Ag^+ under electrospray positive-ion conditions, the peaks corresponding to $[2\text{-}AG+Ag]^+$ ($m/z = 485.2$ and 487.1) predominate, and other commonly observed positive-ion complexes ($[2\text{-}AG+H]^+$, $[2\text{-}AG+18]^+$, and $[2\text{-}AG+Na]^+$) are not seen (Fig. 5.2A). Silver ion coordination results in the observation of two $[M+Ag]^+$ ions because of the natural isotopic distribution of silver (^{107}Ag [52%] and ^{109}Ag [48%]).

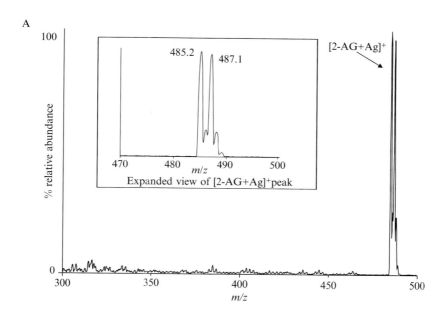

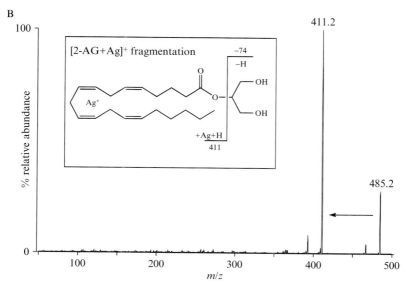

Figure 5.2 (*continued*)

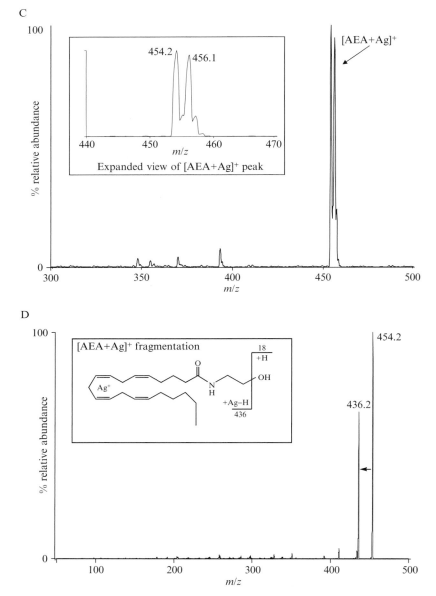

Figure 5.2 Q1 full-scan and CID spectra of 2-AG and AEA, when coordinated with the silver cation. (A) Q1 full-scan spectrum of 2-AG infused with an excess of Ag⁺. Inset shows the double peak resulting from the 107 and 109 silver isotopes. (B) Fragmentation spectrum resulting from CID of $[2\text{-}AG+^{107}Ag]^+$. The inset details the fragmentation site of the analyte complex. (C) Q1 full-scan spectrum of AEA infused with an excess of Ag⁺. (D) Fragmentation spectrum resulting from CID of $[AEA+^{107}Ag]^+$. The inset details the fragmentation site of the analyte complex.

Collision-induced dissociation (CID) of $[2\text{-AG}+^{107}\text{Ag}]^+$ ($m/z = 485.2$) provides an abundant fragment at $m/z = 411.2$ (Fig. 5.2B). Thus, the m/z $485 \rightarrow 411$ transition is used for SRM analysis of 2-AG. Similarly, the m/z $493 \rightarrow 419$ transition is used for the SRM analysis of 2-AG-d_8.

AEA also coordinates strongly with the Ag^+ cation, resulting in a $[\text{AEA}+\text{Ag}]^+$ complex ($m/z = 454.2$ and 456.1; Fig. 5.2C). CID of the $[\text{AEA}+^{107}\text{Ag}]^+$ ($m/z = 454.2$) species gives only one reasonably abundant fragment at $m/z = 436.2$ (Fig. 5.2D). Thus, the m/z $454 \rightarrow 436$ transition is employed for SRM analysis of AEA, while the m/z $462 \rightarrow 444$ transition is used for detection of AEA-d_8. While the loss of 18 amu does not provide greatly enhanced specificity, the combination of silver complexation, chromatographic separation, and tandem mass spectrometric analysis provides sufficient specificity for successful analysis of AEA from mammalian tissue extracts.

One advantage inherent in employing Ag^+ coordination as an ionizing method is that the isotopic distribution of silver provides the analyst a choice of SRM reactions when performing an analysis. For example, if an interfering peak or high background is observed during the analysis of 2-AG, one may switch from m/z $485 \rightarrow 411$ (the $[\text{M}+^{107}\text{Ag}]^+$ ion) to m/z $487 \rightarrow 413$ (the $[\text{M}+^{109}\text{Ag}]^+$ ion). This is also true for AEA analysis. As the ^{107}Ag and ^{109}Ag isotopes are present in roughly equal quantities, there is no difference in sensitivity between SRM reactions. The sensitivity of this method is in the low femtomole on-column range and is comparable to other literature methods. Thus, the isotopic distribution of silver does not hinder the observed limit of detection (LOD).

Figure 5.3 is a representative endocannabinoid chromatogram of processed mouse brain tissue showing the analyte and deuterated internal standard chromatograms. 2-AG exists as a mixture of the 2- and 1(3)-isomers. These isomers are chromatographically resolved with the 2- isomer eluting before the 1- isomer. 2-AG is the initially formed metabolite and undergoes base-catalyzed isomerization to 1(3)-AG with a half-life on the order of 5 to 10 min (Rouzer et al., 2002). The equilibrium composition is approximately 10% 2-AG and 90% 1(3)-AG.

2.3.2. PG-Gs and PG-EAs

Prostaglandins (PGs) deprotonate to generate a strong $[\text{M-H}]^-$ signal in the negative ion mode but PG-Gs and PG-EAs give a very weak signal in negative ion mode (data not shown). However, when infused in the presence of ammonium, PG-Gs and PG-EAs (as well as PGs) undergo complexation with the NH_4^+ cation, resulting in an $[\text{M}+\text{NH}_4]^+$ species. Figure 5.4 shows two examples of this type of ionization. Figure 5.4A depicts mass spectra of PGF$_{2\alpha}$-G-d_5 (MW = 433.6) when infused with ammonium via ESI in the positive ion mode. The dominant peak in the Q1 spectrum (Fig. 5.4A – inset) is m/z 451.2, which corresponds to the $[\text{PGF}_{2\alpha}\text{-G-}d_5+\text{NH}_4]^+$ ion. The larger

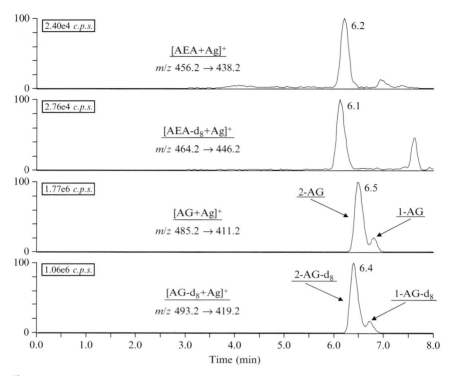

Figure 5.3 Chromatogram of 2-AG, AEA and their octadeuterated internal standards extracted from murine brain tissue. Elution occurred on a C18 column (held at 40°) using a gradient of 70% B to 100% B in 15 min followed by a 10-min hold at 100% B. A = 70 μM silver acetate (aqueous); B = 70 μM in methanol.

spectrum in Fig. 5.4A represents the fragmentation pattern upon CID of m/z 451.2. The loss of NH_3 and water lead to the three main fragments observed: m/z 416.3 ($-NH_3$ and $-H_2O$), 398.5 ($-NH_3$ and $-2H_2O$), and 380.4 ($-NH_3$ and $-3H_2O$).

Figure 5.4B shows the resultant mass spectra when PGD_2-EA (molecular weight = 395.5) is infused in the presence of ammonium via positive ion ESI. Again, the dominant peak is the $[PGD_2\text{-}EA+NH_4]^+$ complex (Fig. 5.4B – inset) with an m/z 413.2. A peak corresponding to $[M+H]^+$ (m/z 396) is also observed but at a significantly lower intensity than the $[M+NH_4]^+$ complex. Upon CID, the $[PGD_2\text{-}EA+NH_4]^+$ ion undergoes loss of ammonia and water to give the major peaks m/z 378.4 ($-NH_3$ and $-H_2O$), 360.4 ($-NH_3$ and $-2H_2O$) and 342.3 ($-NH_3$ and $-3H_2O$). This results in the major fragments seen in the fragmentation spectrum (larger spectrum of Fig. 5.4B). Figure. 5.4 suggests that several ion transitions may be used for the selected reaction monitoring (SRM) of the [prostanoid+NH₄ or +H]$^+$ complexes under consideration. This is indeed the case.

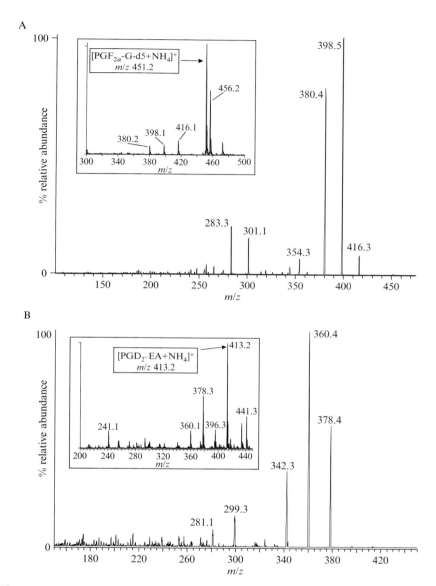

Figure 5.4 Q1 full-scan and CID spectra of $PGF_{2\alpha}$-G-d_5 (A) and PGD_2-EA (B), when infused in the presence of ammonium acetate. (A) Q1 full-scan spectrum of $PGF_{2\alpha}$-G-d_5 infused with an excess of ammonium acetate (inset) and CID spectrum of $[PGF_{2\alpha}$-G-d_5+$NH_4]^+$ (m/z 451.2). (B) Q1 full-scan spectrum of PGD_2-EA infused with ammonium acetate (inset) and CID spectrum of $[PGD_2$-EA+$NH_4]^+$ (m/z 413.2).

For example, in our laboratory, we typically use the m/z 451 → 398 reaction for the SRM monitoring of PGF$_{2\alpha}$-G-d$_5$, but other reactions have been successfully employed for diagnostic or quantitative purposes. Table 5.1 shows the precursor and fragment ions for the SRM detection of the endocannabinoids, PG-Gs, and PG-EAs as used in our laboratory. LODs discussed below were determined using the reactions in Table 5.1; however, this methodology may be easily modified to use alternative reactions. Figure 5.5 displays sample chromatograms of the PG-Gs (Fig. 5.5A) extracted from cell culture medium and a standard mixture of PG-EAs (Fig. 5.5B).

When PGF$_{2\alpha}$-EA and PGE$_2$-EA and their deuterated analogues are infused in the presence of ammonium, they have been observed to ionize by coordination with both the ammonium cation and a proton. However, in contrast to PGD$_2$-EA, the [M+H]$^+$ peak is more abundant than the [M+NH$_4$]$^+$ complex. Thus, the Q1 mass for SRM detection of PGF$_{2\alpha}$-EA and PGE$_2$-EA represents [M+H]$^+$. Of note is the sodiated species [M+Na$^+$] that is seen in the Q1 infusion of most prostanoid analytes (m/z 456.2 in the PGF$_{2\alpha}$-G-d$_5$ infusion discussed above). While the sodiated species produces a strong Q1 signal, CID of this complex provides no strong fragments. Thus, sodium complexation is not useful for SRM analysis.

Table 5.1 Molecular weight and SRM transitions of endocannabinoids and prostanoids

Compound	Molecular weight	SRM transition Q1 m/z → Q3 m/z
2-AG	378.6	485[a] → 411
2-AG-d8	386.6	493[a] → 419
AEA	347.5	454[a] → 436
AEA-d8	355.6	462[a] → 444
PGF$_{2\alpha}$-G	428.6	446[b] → 393
PGF$_{2\alpha}$-G-d5	433.6	451[b] → 398
PGE$_2$-G & PGD$_2$-G	426.6	444[b] → 391
PGE$_2$-G-d5 & PGD$_2$-G-d5	431.6	449[b] → 396
6-keto-PGF$_{1\alpha}$-G	444.6	462[b] → 391
6-keto-PGF$_{1\alpha}$-G-d5	449.6	467[b] → 396
PGF$_{2\alpha}$-EA	397.5	398[c] → 344
PGF$_{2\alpha}$-EA-d4	401.5	402[c] → 348
PGE$_2$-EA	395.5	396[c] → 360
PGE$_2$-EA-d4	399.5	400[c] → 364
PGD$_2$-EA	395.5	413[b] → 360

[a] Q1 m/z represents the [M+^{107}Ag]$^+$ complex.
[b] Q1 m/z represents the [M+NH$_4$]$^+$ complex.
[c] Q1 m/z represents the [M+H]$^+$ complex.

Synthetic preparations of the PG-Gs and their pentadeuterated analogues are a mixture of the 1(3)- and 2-isomers of each species at a 9:1 ratio, respectively (Kozak *et al.*, 2000). PGE$_2$, PGD$_2$, and PGF$_{2\alpha}$ glyceryl esters are chromatographically resolved, with the less abundant 2-isomer eluting before the 1(3)-isomer. It was found that capillary offset, capillary temperature, spray voltage, tube lens offset and sheath, and auxiliary and CID gas pressures optimized at approximately the same values for all analytes. The optimal values for mass spectrometer parameters will change over time and with different instruments. Thus, it is recommended that the settings of any instrument be regularly optimized.

3. QUANTITATION

Quantitation in most literature methods surveyed, as well as in assays performed and reported by this laboratory, is achieved using stable isotope dilution. Stable isotope dilution involves spiking the sample to be processed

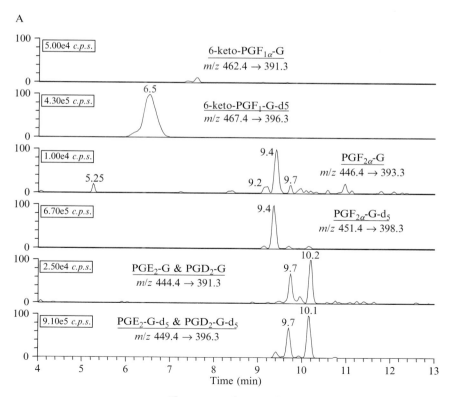

Figure 5.5   (*continued*)

B

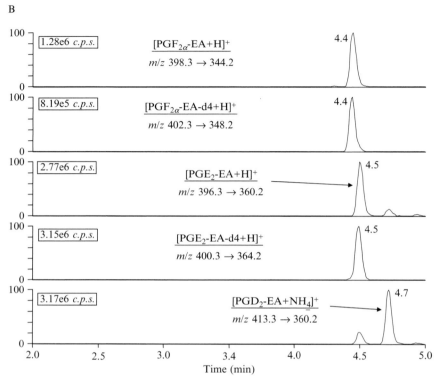

Figure 5.5 Chromatograms of prostanoid species. (A) PG-Gs extracted from RAW 264.7 cell medium. Elution occurred on a C18 column (held at 40°) using a gradient of 25% B to 60% B in 12 min then to 80% B in 2 min. A = 2 mM ammonium acetate, pH 3.3 with 5% acetonitrile and B = 90% acetonitrile and 10% A. (B) A standard mixture of PG-EAs. Elution occurred on a C18 column (held at 40°) using a gradient of 20% B to 50% B in 5.0 min. A = 5 mM ammonium acetate, pH 3.5; B = 90% acetonitrile and 10% A.

with an isotopically enriched internal standard. Typically, the internal standard is a multi–deuterium–labeled analyte, which is commercially available for endocannabinoids and easily synthesized for PG-Gs and PG-EAs (Kozak *et al.*, 2000), using either deuterated prostaglandins or deuterated glycerol or ethanolamine as starting materials.

Quantitation via stable isotope dilution involves transforming the observed response of an analyte (the ratio of the peak area of analyte to that of the internal standard) to amount of analyte based on the known amount of internal standard added to the sample. This is basically a one-point calibration curve and it involves several assumptions. The first assumption is that an analyte and its deuterated internal standard exhibit peak area ratios equivalent to their concentrations. The second assumption is that a deuterated internal standard is "isotopically pure," that is, no undeuterated compound can be found in a preparation of the internal standard. Finally, one must also assume that the analyte and internal standard behave identically throughout any purification process that the sample is subjected to, resulting in identical recoveries.

We have tested the first assumption for the endocannabinoids and the PG-Gs. Solutions of analyte and internal standard were prepared with molar ratios of 1:10, 1:1, and 10:1. The observed ratios can be seen in Table 5.2. The observed ratios for each analyte decrease by a factor of 10 when the molar ratio is changed from 1:1 to 1:10. Likewise, the observed ratios increase 10-fold when the molar ratio is increased from 1:1 to 10:1.

The second assumption was tested merely by injecting a solution containing only internal standard onto the described LC-MS system. One would expect to observe no analyte signal regardless of the size of the analogous internal standard peak. All deuterated internal standards were examined in such a manner and exhibited no signal for their corresponding analyte. Figure 5.6 is a chromatogram from the injection of a solution of PGE_2-G-d_5. The lower transition (m/z 449 → 396) shows a strong signal, while the upper transition (m/z 444 → 391, used to monitor PGE_2-G and PGD_2-G) shows no peaks at the PGE_2-G retention time of 6.9 min. This excellent ratio of internal standard to analyte results from the incorporation of five deuteriums into the internal standard.

Finally, analyte recoveries were evaluated by subjecting samples enriched with either analyte or internal standard to the described purification processes and comparing the resultant LC-MS peak area to that obtained from a sample made at a concentration equal to theoretical 100% recovery. Recovery data can be found in Table 5.3 and are discussed below.

3.1. Endocannabinoid recovery

Recovery of endocannabinoids from an extraction solvent (9:1 ethyl acetate: hexane) was determined by spiking the extraction solvent with 2-AG, 2-AG-d_8, AEA, and AEA-d_8, and following the described procedure to completion.

Table 5.2 Peak area ratios of undeuterated to deuterated analyte

Compound	Theoretical molar ratio (undeuterated:deuterated)		
	1:10	1:1	10:1
2-AG	0.11 ± 0.002	0.97 ± 0.02	9.48 ± 0.05
AEA	0.15 ± 0.002	1.34 ± 0.06	12.2 ± 0.6
$PGF_{2\alpha}$-G	0.079 ± 0.002	0.77 ± 0.024	7.13 ± 0.16
PGE_2-G	0.113 ± 0.02	1.12 ± 0.01	10.7 ± 0.07
PGD_2-G	0.152 ± 0.01	1.51 ± 0.03	13.8 ± 0.11

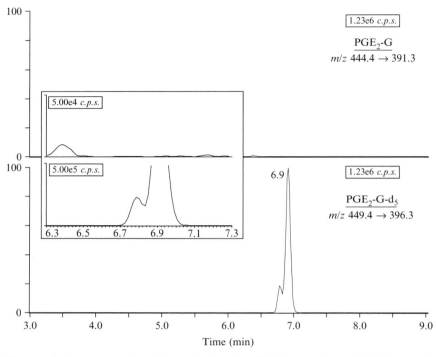

Figure 5.6 Demonstration of isotopical purity of the internal standard PGE$_2$-G-d$_5$. The chromatogram is an injection of PGE$_2$-G-d$_5$ only. The upper transition is PGE$_2$-G, and the lower transition is PGE$_2$-G-d$_5$. The PGE$_2$-G channel is normalized to the intensity of PGE$_2$-G-d$_5$. The inset is a blow-up of the PGE$_2$-G-d$_5$ peak and corresponding time in the PGE$_2$-G transition. No PGE$_2$-G is seen. (See color insert.)

The resultant average percent recovery values ($\pm$ standard deviation) were 73 $\pm$ 5% for 2-AG; 71 $\pm$ 7% for 2-AG-d$_8$; 78 $\pm$ 6% for AEA; and 80 $\pm$ 5% for AEA-d$_8$ ($n = 5$). More than 70% of each analyte is recovered and the recovery of the unlabeled endocannabinoids is similar to the recovery of the octadeuterated internal standards.

Determination of recovery of endocannabinoids from tissue homogenate is complicated by the endogenous nature of these analytes. However, endocannabinoid recovery from tissue homogenates was assessed by enriching murine brain tissue homogenate with 2-AG and AEA. Recoveries were determined at three levels. Table 5.4 gives the results for both analyte and internal standards. The recoveries of the analytes and internal standards were similar to each other and to those observed from extraction solvent only, with the exception of the low-level AEA samples (1 pmol). While these samples exhibited a recovery approximately 30% lower than the 5- and 100-pmol/ml samples and the internal standard, the observed recovery of more than 50% provides reasonable quantitation of AEA from tissue.

Table 5.3   Recoveries of PG-EAs and PG-Gs

Compound	4 pmol/ml	20 pmol/ml	200 pmol/ml	High[a]
PGE$_2$-EA	88 ± 9	77 ± 4	88 ± 6	88 ± 5
PGE$_2$-EA-d4	83 ± 8	74 ± 3	87 ± 6	87 ± 4
PGE$_2$-G	63 ± 4	64 ± 6	65 ± 6	68 ± 3
PGE$_2$-G-d5	63 ± 3	(20 pmol/ml)		
PGD$_2$-EA	79 ± 9	73 ± 4	83 ± 6	86 ± 2
PGD$_2$-G	59 ± 5	55 ± 3	58 ± 4	60 ± 5
PGD$_2$-G-d5	59 ± 3	(20 pmol/ml)		
PGF$_{2\alpha}$-EA	78 ± 11	75 ± 5	88 ± 5	89 ± 5
PGF$_{2\alpha}$-EA-d4	78 ± 10	80 ± 5	84 ± 6	90 ± 7
PGF$_{2\alpha}$-G	64 ± 3	63 ± 3	59 ± 5	74 ± 9
PGF$_{2\alpha}$-G-d5	64 ± 7	(20 pmol/ml)		
6-keto-PGF$_{1\alpha}$-G-d5	76 ± 4	85 ± 2	84 ± 4	76 ± 4

[a] Concentration of this column is 500 pmol/ml for PG–EAs and 1000 pmol/ml for PG–Gs.
Values represent the mean percent recovery ± standard deviation ($n = 5$ for PG–EAs, $n = 4$ for PG–Gs).

Table 5.4   Percent recoveries of endocannabinoids

Compound	1 pmol/ml	5 pmol/ml	100 pmol/ml
AEA	51 ± 4	67 ± 7	82 ± 2
AEA-d8	83 ± 5		
	100 pmol/ml	**500 pmol/ml**	**10,000 pmol/ml**
2-AG	94 ± 79	77 ± 5	76 ± 2
2-AG-d8	83 ± 10		

Values represent the mean percent recovery ± standard deviation ($n = 3$ for analyte, $n = 12$ for internal standards).

3.2. PG-EA and PG-G recovery

Recovery of PG–Gs and PG–EAs was assessed by spiking 1 ml of cell culture medium with prostanoid analytes at four different levels. The analyte concentrations were 4, 20, 200, and 500 (PG–EAs), or 1000 (PG–Gs) pmol/ml. The spiked media samples were purified by the described SPE procedure above and analyzed on the described LC–MS system. Table 5.4 lists the observed percent recovery for each compound.

All prostanoid compounds exhibited recovery greater than 50%, and all compounds showed very similar recoveries across the concentration range. Recoveries were also similar to their deuterated internal standards, if available. This indicates that the analytes and their internal standards respond similarly to the SPE procedure.

The assumptions discussed above have been examined and found to be valid for endocannabinoids and prostanoid species. Thus, stable isotope dilution may be used with confidence and expected to provide reliable quantitation. However, it is recommended that all laboratories interested in using this methodology perform preliminary experiments similar to those described above before sample analysis.

4. LINEARITY AND SENSITIVITY

The linear range of the endocannabinoids and prostanoids under discussion was determined by preparing standard curves and analyzing them on the described systems. Table 5.5 gives the linear range of both endocannabinoids and prostanoids in femtomoles on-column. 2-AG and AEA values were determined on a TSQ 7000 instrument and may not be reflective of the endocannabinoid linearity of a Quantum instrument, which is considerably more sensitive. Also, the linear ranges were chosen to reflect the relative levels of the endocannabinoids *in vivo*, where 2-AG is typically two to three orders of magnitude more abundant than AEA.

The LOD or limit of detection typically represents the smallest amount of material on-column which consistently gives a signal-to-noise (S:N) ratio of greater than 3. LODs of the analytes are also given in Table 5.5. For endocannabinoids, the LOD reported here is comparable to what is reported in other literature methods. The endocannabinoid LOD has not been assessed on a Quantum instrument, and is expected to be lower on that instrument. Additionally, improvements in chromatographic technology, such as the recently introduced UPLC system, may lower LODs (Churchwell *et al.*, 2005).

Table 5.5 Linear range and limit of detection of endocannabinoids and prostanoids

Compound	Linear range (fmol on-column)	Limit of detection (fmol on-column)
2-AG	3300–330000	14
AEA	33–3300	13
$PGF_{2\alpha}$-G	50–25000	5
PGE_2-EA	25–25000	<25
PGE_2-G	25–25000	5
PGD_2-EA	25–25000	<25
PGD_2-G	50–25000	5
6-keto-$PGF_{1\alpha}$-G-d5	5–25000	5

5. Postmortem Changes in Analytes

Postmortem changes in endocannabinoid levels of central nervous system tissue have been investigated in several studies. Kempe *et al.* (1996) and Schmid *et al.* (1995) report significant generation of AEA in brain tissue after decapitation. However, the time scale studied was on the order of hours, and the reported data showed no increase of AEA within seconds of decapitation. Sugiura *et al.* (2001) have reported a rapid postmortem increase in 2-AG levels in rat brain tissue. An increase is seen 15 seconds after decapitation and the maximum 2-AG concentration is achieved at 30 to 60 s after decapitation. This maximum level is roughly five times that of the freshly decapitated brain tissue. An implication of these studies is that postmortem sample handling is a crucial detail and may impact the observed levels of endocannabinoids.

Further complicating the discussion is the issue of microwave fixation. Microwave fixation refers to the practice of exposing the tissue of interest to brief, intense microwave radiation ($\approx$5 kW for up to a few seconds) while the animal is alive and anesthetized. When brain tissue is under consideration, microwave irradiation serves as the method of sacrifice. Bazinet *et al.* (2005) reported that the AEA concentration in murine brain tissue from mice sacrificed by microwave irradiation of 5.5 kW for 3.4 s following anesthetization was four-fold lower than animals sacrificed by decapitation and subjected to microwave irradiation 5 min after sacrifice. Thus, the AEA levels reported in the literature are likely higher than those of a living animal. There is no literature regarding the effect of microwave irradiation on 2-AG levels.

Additionally, there is a significant body of literature demonstrating that microwave treatment of the tissue of interest results in lower levels of both free fatty acids, such as arachidonic acid, and prostaglandins compared to ischemic methods of sacrifice (Anton *et al.*, 1983; Cenedella *et al.*, 1975). Thus, it is reasonable to conclude that, if found to be relevant biological metabolites, PG-G and PG-EA levels would also be affected by ischemic methods of sacrifice. Given the above discussion, there is some uncertainty as to the proper method of sample handling before analysis in order to achieve the most accurate reflection of the basal eicosanoid levels.

6. Summary

Another advantage of the techniques discussed above is their applicability to related compounds not discussed here. For example, non–C20 monoacylglycerol and acylethanolamine species, as well as other endocannabinoids, such as noladin (Hanus *et al.*, 2001), may be detected by the Ag^+ coordination technique. Our lab has conducted preliminary investigations of unsaturated C18 monoacylglycerols and noladin using this method. Additionally, isoprostane glycerol esters and ethanolamides may be analyzed by coordination with the ammonium cation. The limit of detection of the techniques outlined above can be expected to drop as more sensitive triple quadrupole instruments become available and as more sophisticated LC techniques, such as UPLC (ultra pressure liquid chromatography), become more popular and available. Awareness of the importance of lipid research is increasing as the complexity and dynamic nature of the lipidome becomes apparent. The above techniques will allow researchers to take advantage of LC-MS-MS technology, which is now fairly mature and commonplace, to conduct sensitive and specific analyses of neutral eicosanoids.

ACKNOWLEDGMENTS

Supported by the National Institutes of Health (grants GM 15431 and E500267).

REFERENCES

Anton, R. F., Wallis, C., and Randall, C. L. (1983). *In vivo* regional levels of PGE and thromboxane in mouse brain: Effect of decapitation, focused microwave fixation, and indomethacin. *Prostaglandins* **26**, 421–429.

Bazinet, R. P., Lee, H. J., Felder, C. C., Porter, A. C., Rapoport, S. I., and Rosenberger, T. A. (2005). Rapid high-energy microwave fixation is required to determine the anandamide (N-arachidonoylethanolamine) concentration of rat brain. *Neurochem. Res.* **30**, 597–601.

Berdyshev, E. V. (2000). Cannabinoid receptors and the regulation of immune response. *Chem. Phys. Lipids* **108**, 169–190.

Berrendero, F., Sepe, N., Ramos, J. A., Di Marzo, V., and Fernandez-Ruiz, J. J. (1999). Analysis of cannabinoid receptor binding and mRNA expression and endogenous cannabinoid contents in the developing rat brain during late gestation and early postnatal period. *Synapse* **33**, 181–191.

Canty, A. J., and Colton, R. (1994). P-Bonded alkene and arene complexes of silver(I): An electrospray mass spectrometric study. *Inorg. Chim. Acta* **220**, 99–105.

Cenedella, R. J., Galli, C., and Paoletti, R. (1975). Brain free fatty levels in rats sacrificed by decapitation versus focused microwave irradiation. *Lipids* **10**, 290–293.

Churchwell, M. I., Twaddle, N. C., Meeker, L. R., and Doerge, D. R. (2005). Improving LC-MS sensitivity through increases in chromatographic performance: Comparisons of UPLC-ES/MS/MS to HPLC-ES/MS/MS. *J. Chromatogr. B Anal. Technol. Biomed. Life Sci.* **825**, 134–143.

Devane, W. A., Hanus, L., Breuer, A., Pertwee, R. G., Stevenson, L. A., Griffin, G., Gibson, D., Mandelbaum, A., Etinger, A., and Mechoulam, R. (1992). Isolation and structure of a brain constituent that binds to the cannabinoid receptor. *Science* **258**, 1946–1949.

Di Marzo, V., Goparaju, S. K., Wang, L., Liu, J., Batkai, S., Jarai, Z., Fezza, F., Miura, G. I., Palmiter, R. D., Sugiura, T., and Kunos, G. (2001). Leptin-regulated endocannabinoids are involved in maintaining food intake. *Nature* **410**, 822–825.

Hanus, L., Abu-Lafi, S., Fride, E., Breuer, A., Vogel, Z., Shalev, D. E., Kustanovich, I., and Mechoulam, R. (2001). 2-arachidonyl glyceryl ether, an endogenous agonist of the cannabinoid CB1 receptor. *Proc. Natl. Acad. Sci. USA* **98**, 3662–3665.

Havrilla, C. M., Hachey, D. L., and Porter, N. (2000). Coordination (Ag+) ion spray-mass spectrometry of peroxidation products of cholesterol linoleate and cholesterol arachidonate: High-performance liquid chromatography-mass spectrometry analysis of peroxide products from polyunsaturated lipid autoxidation. *J. Am. Chem. Soc.* **122**, 8042–8055.

Kempe, K., Hsu, F. F., Bohrer, A., and Turk, J. (1996). Isotope dilution mass spectrometric measurements indicate that arachidonylethanolamide, the proposed endogenous ligand of the cannabinoid receptor, accumulates in rat brain tissue post mortem but is contained at low levels in or is absent from fresh tissue. *J. Biol. Chem.* **271**, 17287–17295.

Kingsley, P. J., and Marnett, L. J. (2003). Analysis of endocannabinoids by Ag+ coordination tandem mass spectrometry. *Anal. Biochem.* **314**, 8–15.

Kingsley, P. J., Rouzer, C. A., Saleh, S., and Marnett, L. J. (2005). Simultaneous analysis of prostaglandin glyceryl esters and prostaglandins by electrospray tandem mass spectrometry. *Anal. Biochem.* **343**, 203–211.

Koga, D., Santa, T., Fukushima, T., Homma, H., and Imai, K. (1997). Liquid chromatographic-atmospheric pressure chemical ionization mass spectrometric determination of anandamide and its analogs in rat brain and peripheral tissues. *J. Chromatogr. B Biomed. Sci. Appl.* **690**, 7–13.

Kondo, S., Kondo, H., Nakane, S., Kodaka, T., Tokumura, A., Waku, K., and Sugiura, T. (1998). 2-Arachidonoylglycerol, an endogenous cannabinoid receptor agonist: Identification as one of the major species of monoacylglycerols in various rat tissues, and evidence for its generation through $CA2^+$-dependent and -independent mechanisms. *FEBS Lett.* **429**, 152–156.

Kozak, K. R., Rowlinson, S. W., and Marnett, L. J. (2000). Oxygenation of the endocannabinoid, 2-arachidonylglycerol, to glyceryl prostaglandins by cyclooxygenase-2. *J. Biol. Chem.* **275**, 33744–33749.

Kozak, K. R., Crews, B. C., Morrow, J. D., Wang, L. H., Ma, Y. H., Weinander, R., Jakobsson, P. J., and Marnett, L. J. (2002). Metabolism of the endocannabinoids, 2-arachidonylglycerol and anandamide, into prostaglandin, thromboxane, and prostacyclin glycerol esters and ethanolamides. *J. Biol. Chem.* **277**, 44877–44885.

Maccarrone, M., Attina, M., Cartoni, A., Bari, M., and Finazzi-Agro, A. (2001). Gas chromatography-mass spectrometry analysis of endogenous cannabinoids in healthy and tumoral human brain and human cells in culture. *J. Neurochem.* **76**, 594–601.

Mechoulam, R., Ben-Shabat, S., Hanus, L., Ligumsky, M., Kaminski, N. E., Schatz, A. R., Gopher, A., Almog, S., Martin, B. R., Compton, D. R., Pertwee, R. G., Griffin, G., et al. (1995). Identification of an endogenous 2-monoglyceride, present in canine gut, that binds to cannabinoid receptors. *Biochem. Pharmacol.* **50**, 83–90.

Morris, L. J. (1966). Separations of lipids by silver ion chromatography. *J. Lipid Res.* **7**, 717–732.

Nichols, P. L. (1952). Coordination of silver ion with methyl esters of oleic and elaidic acids. *J. Am. Chem. Soc.* **74**, 1091–1093.

Nirodi, C. S., Crews, B. C., Kozak, K. R., Morrow, J. D., and Marnett, L. J. (2004). The glyceryl ester of prostaglandin E2 mobilizes calcium and activates signal transduction in RAW264.7 cells. *Proc. Natl. Acad. Sci. USA* **101**, 1840–1845.

Rouzer, C. A., Ghebreselasie, K., and Marnett, L. J. (2002). Chemical stability of 2-arachidonylglycerol under biological conditions. *Chem. Phys. Lipids* **119**, 69–82.

Rouzer, C. A., and Marnett, L. J. (2005). Glycerylprostaglandin synthesis by resident peritoneal macrophages in response to a zymosan stimulus. *J. Biol. Chem.* **280,** 26690–26700.

Schmid, P. C., Krebsbach, R. J., Perry, S. R., Dettmer, T. M., Maasson, J. L., and Schmid, H. H. (1995). Occurrence and postmortem generation of anandamide and other long-chain N-acylethanolamines in mammalian brain. *FEBS Lett.* **375,** 117–120.

Schmid, P. C., Schwartz, K. D., Smith, C. N., Krebsbach, R. J., Berdyshev, E. V., and Schmid, H. H. (2000). A sensitive endocannabinoid assay. The simultaneous analysis of N-acylethanolamines and 2-monoacylglycerols. *Chem. Phys. Lipids* **104,** 185–191.

Sugiura, T., Kondo, S., Sukagawa, A., Nakane, S., Shinoda, A., Itoh, K., Yamashita, A., and Waku, K. (1995). 2-Arachidonoylglycerol: A possible endogenous cannabinoid receptor ligand in brain. *Biochem. Biophys. Res. Commun.* **215,** 89–97.

Sugiura, T., Yoshinaga, N., and Waku, K. (2001). Rapid generation of 2-arachidonoylglycerol, an endogenous cannabinoid receptor ligand, in rat brain after decapitation. *Neurosci. Lett.* **297,** 175–178.

Van Sickle, M. D., Duncan, M., Kingsley, P. J., Mouihate, A., Urbani, P., Mackie, K., Stella, N., Makriyannis, A., Piomelli, D., Davison, J. S., Marnett, L. J., Di Marzo, V., *et al.* (2005). Identification and functional characterization of brainstem cannabinoid CB2 receptors. *Science* **310,** 329–332.

Weber, A., Ni, J., Ling, K. H., Acheampong, A., Tang-Liu, D. D., Burk, R., Cravatt, B. F., and Woodward, D. (2004). Formation of prostamides from anandamide in FAAH knockout mice analyzed by HPLC with tandem mass spectrometry. *J. Lipid Res.* **45,** 757–763.

Wilson, R. I., and Nicoll, R. A. (2001). Endogenous cannabinoids mediate retrograde signalling at hippocampal synapses. *Nature* **410,** 588–592.

Winstein, S., and Lucas, H. J. (1938). The coordination of silver ion with unsaturated compounds. *J. Am. Chem. Soc.* **60,** 836.

Yu, M., Ives, D., and Ramesha, C. S. (1997). Synthesis of prostaglandin E2 ethanolamide from anandamide by cyclooxygenase-2. *J. Biol. Chem.* **272,** 21181–21186.

QUANTIFICATION OF F2-ISOPROSTANES IN BIOLOGICAL FLUIDS AND TISSUES AS A MEASURE OF OXIDANT STRESS

Ginger L. Milne, Huiyong Yin, Joshua D. Brooks, Stephanie Sanchez, L. Jackson Roberts, II, *and* Jason D. Morrow

Contents

Abstract

Oxidant stress has been implicated in a wide variety of disease processes. One method to quantify oxidative injury is to measure lipid peroxidation. Quantification of a group of prostaglandin F_2-like compounds derived from the nonezymatic oxidation of arachidonic acid, termed the F_2-isoprostanes (F_2-IsoPs), provides an accurate assessment of oxidative stress both *in vitro* and *in vivo*. In fact, in a recent National Institutes of Health–sponsored independent study, F_2-IsoPs were shown to be the most reliable index of *in vivo* oxidant stress when compared against other well-known biomarkers. This article summarizes current methodology used to quantify these molecules. Our laboratory's method to measure F_2-IsoPs in biological fluids and tissues using gas chromatography-mass spectrometry

Division of Clinical Pharmacology, Vanderbilt University School of Medicine, Nashville, Tennessee

Methods in Enzymology, Volume 433
ISSN 0076-6879, DOI: 10.1016/S0076-6879(07)33006-1

is detailed herein. In addition, other mass spectrometric approaches, as well as immunological methods to measure these compounds, are discussed. Finally, the utility of these molecules as *in vivo* biomarkers of oxidative stress is summarized.

1. INTRODUCTION

Free radicals, largely derived from molecular oxygen, have been implicated in a variety of human conditions and diseases, including atherosclerosis and associated risk factors, cancer, neurodegenerative diseases, and aging. Damage to tissue biomolecules by free radicals is postulated to contribute importantly to the pathophysiology of oxidative stress (Chisholm and Steinberg, 2000; Halliwell and Gutteridge, 1990). Measuring oxidative stress in humans requires accurate quantification of either free radicals or damaged biomolecules. The targets of free radical-mediated oxidant injury include lipids, proteins, and DNA. A number of methods exist to quantify free radicals and their oxidation products, although many of these techniques suffer from lack of sensitivity and specificity, especially when used to assess oxidant stress status *in vivo*. In a recent multi-investigator study, termed the Biomarkers of Oxidative Stress Study (BOSS), sponsored by the National Institutes of Health, it was found that the most accurate method to assess *in vivo* oxidant stress status is the quantification of plasma or urinary isoprostanes (IsoPs; Kadiiska *et al.*, 2005). IsoPs, a series of prostaglandin (PG)-like compounds produced by the free radical–catalyzed peroxidation of arachidonic acid independent of the cyclooxygenase, were first discovered by our laboratory in 1990 (Morrow *et al.*, 1990a). The mechanism of formation of these compounds has been intensely studied and is reviewed in the literature (Famm and Morrow, 2003; Morrow *et al.*, 1999a).

F_2-IsoPs (structure shown in Fig. 6.1) are stable, robust molecules and are detectable in all human tissues and biological fluids analyzed, including

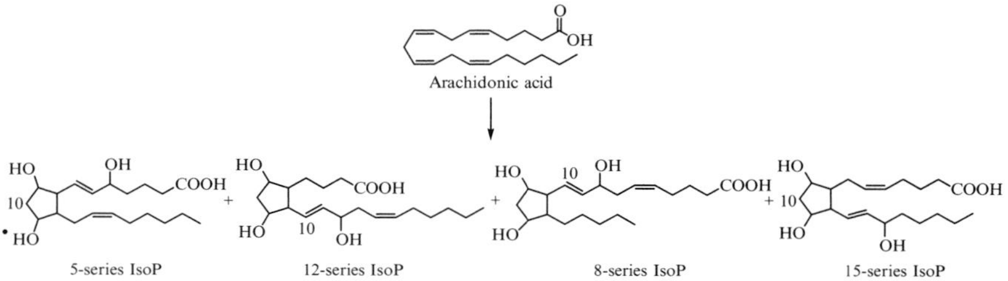

Figure 6.1   Structure of F_2-isoprostanes derived from the oxidation of arachidonic acid.

plasma, urine, bronchoalveolar lavage fluid, cerebrospinal fluid, and bile (Morrow et al., 1999a). The quantification of F_2-IsoPs in urine and plasma, however, is most convenient and least invasive. Based on the available data, quantification of these compounds in either plasma or urine is representative of their endogenous production, and thus gives a highly precise and accurate index of in vivo oxidant stress.

Normal levels of F_2-IsoPs in healthy humans have been defined (Liang et al., 2003; Morrow and Roberts, 1999; Morrow et al., 1999a). Defining these levels is particularly important in that it allows for an assessment of the effects of diseases on endogenous oxidant tone and permits the determination of the extent to which various therapeutic interventions affect levels of oxidant stress. Elevations of IsoPs in human body fluids and tissues have been found in a diverse array of human disorders, some of which include atherosclerosis, hypercholesterolemia, diabetes, obesity, cigarette smoking, neurodegenerative diseases, rheumatoid arthritis, and many others. Further, treatments for some of these conditions, including antioxidant supplementation, antidiabetic treatments, cessation of smoking, and even weight loss, have been shown to decrease production of F_2-IsoPs. Thus, the clinical utility of F_2-IsoPs has been great and continues to grow.

Several methods have been developed to quantify the F_2-IsoPs from biological sources. Our laboratory uses a gas chromatographic/negative ion chemical ionization mass spectrometric (GC/NICI-MS) approach employing stable isotope dilution that will be detailed herein (Milne et al., 2007; Morrow et al., 1999a). Specific examples will be given demonstrating both the utility and limitations of the assay. Further, procedures will be outlined for the analysis of both free F_2-IsoPs and F_2-IsoPs esterified to phospholipids. A number of alternative methods to quantify F_2-IsoPs utilizing both mass spectrometric and immunological approaches have been developed by different investigators and will be reviewed at the end of this chapter.

For quantification purposes, we measure the F_2-IsoP, 15-F_{2t}-IsoP, and other F_2-IsoPs that co-elute on GC with this compound. The advantages of this technique over other approaches include its high sensitivity and specificity, which yields quantitative results in the low picogram range. Its drawbacks are that it is labor intensive and requires considerable expenditures on equipment.

2. HANDLING AND STORAGE OF BIOLOGICAL FLUIDS AND TISSUES FOR QUANTIFICATION OF F_2-ISOPS

As discussed, F_2-IsoPs have been detected in all biological fluids and tissues examined thus far. A potential drawback to measuring F_2-IsoPs as an index of endogenous lipid peroxidation is that they can be readily generated *ex vivo* in biological fluids, such as plasma, in which arachidonoyl-containing lipids are present. This occurs not only when biological fluids or tissues are left at room temperature but also occurs if they are stored at $-20°$ (Morrow *et al.*, 1990a). However, we have found that the formation of F_2-IsoPs does not occur if biological fluids or tissues are processed immediately after procurement, and if the agents including butylated hydroxytoluene (BHT, a free radical scavenger) and/or triphenylphosphine (a reducing agent) are added to the organic solvents during extraction and hydrolysis of phospholipids (Morrow and Roberts, 1999; Morrow *et al.*, 1990b). Thus, samples to be analyzed for F_2-IsoPs should either be processed immediately or stored at $-70°$. Ideally, samples should be rapidly frozen in liquid nitrogen prior to placement at $-70°$. This is especially important for tissue samples, and we routinely snap freeze them with a clamp before storage at $-70°$, because it is known that inner areas of tissue samples that are not snap frozen may remain in a liquid state for a period of time, even when placed at $-70°$.

3. EXTRACTION AND HYDROLYSIS OF F_2-IsoP–CONTAINING PHOSPHOLIPIDS IN TISSUES AND BIOLOGICAL FLUIDS

To measure levels of F_2-IsoPs esterified in tissue phospholipids, the phospholipids must first be extracted from the tissue sample and subjected to alkaline hydrolysis to release free F_2-IsoPs. Free F_2-IsoPs are then quantified using the same procedure for the measurement of free compounds in biological fluids. Therefore, the procedure to extract and hydrolyze F_2-IsoPs from tissue lipids will be outlined first. Subsequently, the method of analysis for free compounds will be discussed.

To 0.05 to 0.5 g of tissue is added 20 ml of ice-cold Folch solution, $CHCl_3$/methanol (2:1, v/v), containing 0.005% BHT in a 50 ml centrifuge tube. As discussed, the presence of BHT during extraction and hydrolysis of lipids is important because it completely inhibits *ex vivo* formation of F_2-IsoPs during this procedure (Morrow and Roberts, 1999; Morrow *et al.*, 1990a). The tissue is then homogenized with a blade homogenizer for 30 s, and the mixture is allowed to stand sealed under nitrogen at room temperature for 1 h, shaking occasionally. Four milliliters of aqueous NaCl (0.9%) are then added and the solution is vortexed vigorously and centrifuged at $800 \times g$ for 10 min. After centrifugation, the upper aqueous layer is discarded and the lower organic layer is carefully separated from the intermediate semisolid proteinaceous layer.

The organic phase containing the extracted lipids is then transferred to a second 50-ml centrifuge tube and evaporated to dryness under a stream of N_2. Two to four milliliters of methanol containing BHT (0.005%) and an equal volume of aqueous KOH (15%) are then added to the residue. The mixture is then vortexed and incubated at 37° for 30 min to effect hydrolysis and release of the F_2-IsoPs. The mixture is then acidified to pH 3 with 1 M HCl and diluted to a final volume of at least 40 ml with pH 3 water in preparation for extraction of free F_2-IsoPs (see discussion below). Dilution of the methanol in the solution with water to 5% or less is necessary to ensure proper column extraction of F_2-IsoPs in the subsequent purification procedure.

For the extraction of lipids containing esterified F_2-IsoPs in biological fluids such as plasma, as opposed to tissue, a different method is used (Milne *et al.*, 2007; Morrow and Roberts, 1999). We have found that the addition of BHT alone to the Folch solution used to extract plasma lipids does not suppress *ex vivo* lipid peroxidation entirely. On the other hand, the addition of both triphenylphosphine and BHT to the organic solvents used to extract plasma lipids entirely prevents autoxidation. Thus, 20 ml of ice-cold Folch solution, and $CHCl_3$/methanol (2:1, v/v), containing 0.005% BHT and triphenylphosphine (5 mg) are added to 1 ml of a biological fluid such as plasma in a 40-ml centrifuge tube. After shaking the mixture for 2 min, 10 ml of 0.043% aqueous $MgCl_2$ are added, followed by shaking for 2 min and then centrifugation at $800 \times g$ for 10 min. Next, the upper aqueous layer is discarded and the lower organic layer is carefully separated from the intermediate semisolid proteinaceous layer. The organic layer is then dried under nitrogen and subsequently subjected to hydrolysis using the same procedure outlined above for tissue lipids.

4. PURIFICATION, DERIVATIZATION, AND QUANTIFICATION OF FREE F_2-IsoPs

Quantification of F_2-IsoPs by GC/NICI-MS is extremely sensitive, with a lower limit of detection in the range of 1 to 5 pg using a deuterated internal standard with a $[^2H_0]$ blank of less than 5 parts per thousand. Thus, it is usually not necessary to assay more than 1 to 3 ml of a biological fluid or a lipid extract from more than 50 to 100 mg of tissue. Further, because urinary levels of F_2-IsoPs are high (typically greater than 1 ng/ml), we have found that 0.25 ml of urine is more than adequate to quantify urinary F_2-IsoPs. The following assay procedure, described and summarized in Fig. 6.2, is the method used for analysis of free F_2-IsoPs in plasma, but is equally adaptable for quantifying F_2-IsoPs in urine, other biological fluids such as cerebral spinal fluid (CSF), and hydrolyzed lipid extracts of tissues.

Following acidification of 3 ml of plasma to pH 3 with 1 M HCl, 1 ng of the deuterated internal standard $[^2H_4]$-15-F_{2t}-IsoP (8–iso-PGF$_{2\alpha}$; Cayman Chemical, Ann Arbor, MI) is added. After addition of the internal standard, the mixture is vortexed and applied to a C_{18} Sep-Pak column (Waters

Biological fluid or hydrolyzed lipid extract;
acidify to pH 3; add deuterated standard ($[^2H_4]$-15-F_{2t}-IsoP)

↓

C-18 and silica Sep-Pak extraction

↓

Formation of PFB esters

↓

TLC of F_2-IsoPs as PFB esters

↓

Formation of trimethylsilyl ether derivatives

↓

Quantification by selected ion monitoring GC/NICI-MS

Figure 6.2   Outline of the procedures used for the extraction, purification, derivatization, and mass spectrometric analysis of F_2-IsoPs from biological sources.

Associates, Milford, MA) preconditioned with 5 ml methanol and 5 ml of water (pH 3). The sample and subsequent solvents are eluted through the Sep-Pak using a 10-ml sterile plastic syringe. The column is then washed sequentially with 10 ml of water (pH 3) and 10 ml of heptane. The F_2-IsoPs are eluted with 10 ml of ethyl acetate/heptane (50/50, v/v).

The ethyl acetate/heptane eluate from the C_{18} Sep-Pak is then dried over anhydrous Na_2SO_4 and applied to a silica Sep-Pak (Waters Associates) that was prewashed with 5 ml of ethyl acetate. The cartridge is then washed with 5 ml of ethyl acetate followed by elution of the F_2-IsoPs with 5 ml of ethyl acetate/methanol (50:50 v/v). The ethyl acetate/methanol eluate is evaporated under a stream of nitrogen.

The F_2-IsoPs are then converted to pentafluorobenzyl (PFB) esters by treatment with a mixture of 40 μl of 10% pentafluorobenzyl bromide in acetonitrile and 20 μl of 10% N,N-diisopropylethylamine in acetonitrile at room temperature for 30 min. The reagents are then dried under nitrogen. The residue is dissolved in 50 μl chloroform/methanol (3/2, v/v) and subjected to thin-layer chromatography (TLC) using the solvent chloroform/ethanol (93:7 v/v). As a TLC standard, approximately 2 to 5 μg of the methyl ester of $PGF_{2\alpha}$ are chromatographed on a separate plate simultaneously. (The methyl ester of $PGF_{2\alpha}$, rather than the PFB ester, is used as the TLC standard because any contamination of the sample being quantified with the methyl ester of $PGF_{2\alpha}$ will not interfere with the analysis owing to the fact that the F_2-IsoPs are analyzed as PFB esters.) After chromatography, the standard is visualized by spraying the plate with a 10% solution of phosphomolybdic acid in ethanol followed by heating. Compounds migrating in the region of the methyl ester of $PGF_{2\alpha}$ ($R_f \sim 0.15$) and the adjacent areas 1 cm above and below are scraped and extracted from the silica gel with ethyl acetate.

The ethyl acetate is dried under a stream of nitrogen, and the F_2-IsoPs are then converted to trimethylsilyl ether derivatives by adding 20 μl of N, O-bis-(trimethylsilyl)trifluoroacetamide (BSTFA) and 10 μl of dimethyl-formamide and incubating at 40° for 20 min. The reagents are dried under a stream of nitrogen and the derivatized F_2-IsoPs are redissolved in 10 μl of undecane, which has been dried over calcium hydride, for analysis by GC/MS.

For quantification of F_2-IsoPs by GC/MS, we routinely use an Agilent 5973 mass spectrometer with a computer interface, although other mass spectrometers can be utilized. The F_2-IsoPs are chromatographed on a 15-m DB1701 fused silica capillary column because we have found that it gives excellent separation of individual regioisomers compared to other columns. The column temperature is programmed from 190° to 300° at 20° per minute. Methane is used as the carrier gas for NICI. The ion source temperature is 200°. The ion monitored for endogenous F_2-IsoPs is the carboxylate anion m/z 569 (M–181, loss of $CH_2C_6F_5$). The corresponding carboxylate anion for the deuterated internal standard is m/z 573. The sensitivity of the mass spectrometer is checked daily by injecting a standard consisting of 40 pg each of $PGF_{2\alpha}$ and $[^2H_4]$-15-F_{2t}-IsoP.

5. Application of the Assay to Quantify F_2-IsoPs in Biological Tissues and Fluids

We have successfully employed this assay to quantify F_2-IsoPs in a number of diverse biological fluids and tissues. Shown in Fig. 6.3 is a selected ion monitoring (SIM) chromatogram obtained from the analysis of F_2-IsoPs in rat plasma. The series of peaks in the upper m/z 569 selected ion current chromatogram represents different endogenous F_2-IsoPs. This pattern of peaks is virtually identical to that obtained from all other biological fluids and tissues that we have examined to date. In the lower m/z 573 chromatogram, the single peak represent the $[^2H_4]$-15-F_{2t}-IsoP internal standard that was added to the plasma sample. For quantification purposes, the peak denoted by the star (*), which co-elutes with the 15-F_{2t}-IsoP internal standard, is routinely measured. Using the ratio of the intensity of this peak to that of the internal standard, the concentration of F_2-IsoPs was calculated to 83 pg/ml. Normal concentrations of F_2-IsoPs in human plasma have been calculated to be 35 + 6 pg/ml (mean plus one 1 standard deviation) while normal concentrations in human urine are 1.6 + 0.6 ng/ml (mean plus one standard deviation). The reader is referred to previously

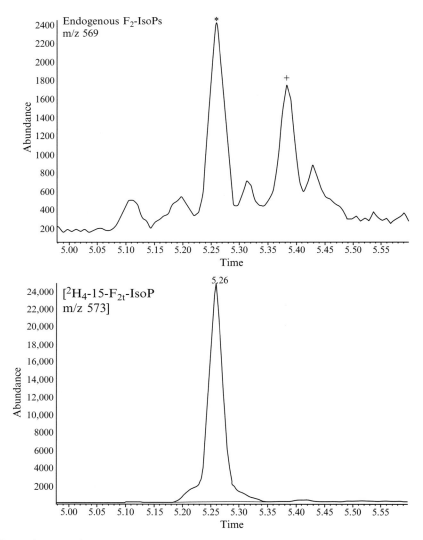

Figure 6.3 Analysis of F_2-IsoPs in plasma obtained from a rat 4 h after treatment with CCl_4 (2 ml/kg orogastrically) to induce endogenous lipid peroxidation. The *m/z* 573 ion current chromatogram represent the [$^2H_4$] 8-iso-PGF$_{2\alpha}$ internal standard. The *m/z* 569 ion current chromatogram represents endogenous F_2-IsoPs. The peak in the upper chromatogram represented by the star (*) is the one routinely used for quantification of the F_2-IsoPs. The peak represented by the plus (+) can be comprised of both F_2-IsoPs and cyclooxygenase-derived PGF$_{2\alpha}$. The concentration of F_2-IsoPs in the plasma using the starred (*) peak for quantification was calculated to be 83 pg/ml.

published data for normal levels of F_2-IsoPs in other biological fluids and tissues (Morrow and Roberts, 1999).

Quantification of the F_2-IsoPs based on the intensity of the starred (*) peak shown in Fig. 6.3 is highly precise and accurate. The precision is +6% and accuracy is 96%. Accuracy was determined by quantification of an added known amount of $[^2H_0]$-15-F_{2t}-IsoP. It should be noted herein that $PGF_{2\alpha}$, which is derived from primarily from cyclooxygenase, does not co-chromatograph with the peak being quantified by this assay. Rather, it elutes with the peak noted by a cross (+). The intensity of this peak varies, as $PGF_{2\alpha}$ is the result of an enzymatic process rather than oxidant stress. Therefore, we have chosen to only quantify the starred peak in our analysis and not subsequently eluting peaks.

6. ALTERNATIVE METHODS FOR MEASURING F_2-ISOPS

It should be noted that several alternative GC/MS assays have been developed by different investigators, including FitzGerald et al. (Pratico et al., 1998; Rokach et al., 1997a, b). Like our assay, these methods quantify F_2-IsoPs using stable isotope-dilution GC/NICI-MS and require solid-phase extraction using a C-18 column, TLC purification, and chemical derivatization. These assays, however, measure F_2-IsoP isomers other than 15-F_2-IsoP, but are comparable to ours in terms of utility.

In addition to these GC/MS assays, a number of liquid chromatographic-MS (LC/MS) methods for F_2-IsoPs have been developed. One advantage of LC/MS methods is that the sample preparation for analysis is simpler than that for GC/MS because no derivatization of the molecule is required. Despite major advancements in the sensitivity of LC/MS instrumentation, a concern with these assays relates to the limits of detection in biological fluids that are often higher than those employing GC/MS. A number of the more recently reported methodologies, however, do offer particular advantages. For example, Zhang and Saku (2007) were able to separate a number of different F_2-IsoP stereoisomers employing novel HPLC gradients that would allow the selective and simultaneous measurement of these compounds. Further, these authors developed a multidimensional solid-phase extraction technique for urine sample clean-up that minimized matrix-related ion suppression effects associated with LC/MS analysis. In another recent report, Masoodi and Nicolaou (2006) developed a LC/MS lipidomic method to simultaneously assay formation of 27 different prostanoids, including F_2-IsoPs. Importantly, however, of the many published methodologies for measuring IsoPs by LC/MS, only three of the most recently published reports have been validated for quantitation of these molecules in biological fluids (Haschke *et al.*, 2007; Sircar and Subbaiah, 2007; Taylor *et al.*, 2006).

Alternative methods have also been developed to quantify IsoPs using immunological approaches. Antibodies have been generated against 15-F_{2t}-IsoP, and at least three immunoassay kits are commercially available. While mass spectrometric methods of IsoP quantification are considered the best methods for analysis, immunoassays have expanded research in this area because of their low cost and relative ease of use. Only limited information is currently available regarding the precision and accuracy of immunoassays. In addition, little data exist comparing IsoP levels determined by immunoassay to MS. Although Wang *et al.* (1995) offer one example of a MS-validated immunoassay to measure urinary F_2-IsoPs, our laboratory's own experiments have not validated the commercially available kits. Analogous to immunological methods to quantify cyclooxygenase-derived PGs, immunoassays for IsoPs suffer from a lack of specificity. Furthermore, the sensitivity and/or specificity of these kits vary substantially among manufacturers.

7. F_2-Isoprostanes as an Index of Oxidant Stress *In Vivo*

The true utility of the F_2-IsoPs is in the quantification of lipid peroxidation and thus oxidant stress status *in vivo*. As discussed, F_2-IsoPs are stable, robust molecules, and are detectable in all human tissues and biological fluids analyzed, including plasma, urine, bronchoalveolar lavage fluid, cerebrospinal fluid, and bile (Morrow *et al.*, 1999a). Based on available data, quantification of these compounds in either plasma or urine is representative of their endogenous production, and thus gives a highly precise and accurate index of *in vivo* oxidant stress. While measurement of F_2-IsoPs in plasma is indicative of their endogenous formation at a specific point in time, analysis of these compounds in urine is an index of systemic or "whole-body" oxidant stress integrated over time. It is important to note that the measurement of free F_2-IsoPs in urine may be confounded by the potential contribution of local IsoP production in the kidney, although this has not been proven to be the case definitively. In light of this issue, we have identified the primary urinary metabolite of 15-F_{2t}-IsoP to be 2,3-dinor-5,6-dihydro-15-F_{2t}-IsoP (Roberts *et al.*, 1996), and we developed a highly sensitive and accurate GC/MS assay to quantify this molecule (Morrow *et al.*, 1999b). Recently, Davies *et al.* (2006) reported a rapid method by which dinor, dihydro-F_2-IsoP metabolites can be measured using LC/MS without previous derivatization. However, the extent to which F_2-IsoPs are converted to the corresponding urinary metabolites remains unclear. Nevertheless, the quantification of 2,3-dinor-5,6-dihydro-15-F_{2t}-IsoP and other dinor,dihydro-F_2-IsoP metabolites may represent a noninvasive, time-integrated measurement of systemic oxidation status that can be applied to living subjects. At present, however, it is our opinion that the quantification of either F_2-IsoPs or their metabolites in urine provides an equally useful systemic index of oxidant stress.

Normal levels of F_2-IsoPs in healthy humans have been defined (Liang et al., 2003; Morrow and Roberts, 1999; Morrow et al., 1999a). Defining these levels is particularly important in that it allows for an assessment of the effects of disease on endogenous oxidant tone and allows for the determination of the extent to which various therapeutic interventions affect levels of oxidant stress. Elevations of IsoPs in human body fluids and tissues have been found in a diverse array of human disorders, some of which include atherosclerosis, hypercholesterolemia, diabetes, obesity, cigarette smoking, neurodegenerative diseases, rheumatoid arthritis, and many others (Basu et al., 2001; Davi et al., 1997, 2002; Gniwotta et al., 1997; Gopaul et al., 1995; Gross et al., 2005; Keaney et al., 2003; Milne et al., 2005; Montine et al., 1998, 1999, 2004; Morrow, 2005; Morrow et al., 1995). Treatments for some of these conditions, including antioxidant supplementation, antidiabetic treaments, cessation of smoking, and even weight loss, have been shown to decrease production of F_2-IsoPs (Davi et al., 2002; Dietrich et al., 2002; Morrow, 2005). Thus, the clinical utility of F_2-IsoPs has been great and continues to grow.

8. SUMMARY

This chapter has outlined methods to assess lipid peroxidation associated with oxidant injury in vivo by quantifying concentrations of free F_2-IsoPs in biological fluids and levels of F_2-IsoPs esterifed in tissue lipids. Our GC/MS assay described herein is highly precise and accurate. A potential shortcoming with this approach is that it requires expensive instrumentation, that is, a mass spectrometer. However, several immunoassays for the F_2-IsoP, 8-iso-PGF$_{2\alpha}$, have become available from commercial sources. At this time, the accuracy and reliability of these assays for quantifying F_2-IsoPs in biological fluids have not been fully validated by mass spectrometry. If these immunoassays prove to be a reliable measure of F_2-IsoPs, however, this should greatly expand the use of F_2-IsoPs to assess oxidant stress. In conclusion, studies carried out over the past several years have shown that measurement of F_2-IsoPs has overcome many of the limitations associated with other methods to assess oxidant status, especially when applied to the measurement of oxidant stress in vivo in humans. Therefore, the quantification of F_2-IsoPs represents an important advance in our ability to assess the role of oxidant stress and lipid peroxidation in human disease.

ACKNOWLEDGMENTS

Supported by National Institutes of Health grants DK48831, ES000267, ES13125, GM42056, and GM15431.

REFERENCES

Basu, S., Whiteman, M., Mattey, D. L., and Halliwell, B. (2001). Raised levels of F(2)-isoprostanes and prostaglandin F(2alpha) in different rheumatic diseases. *Ann. Rheum. Dis.* **60,** 627–631.

Chisholm, G. M., and Steinberg, D. (2000). The oxidative modification hypothesis of atherogenesis: An overview. *Free Radic. Biol. Med.* **18,** 1815–1826.

Davi, G., Alessandrini, P., Mezzetti, A., Minotti, G., Bucciarelli, T., Costantini, F., Cipollone, F., Bon, G. B., Ciabattoni, G., and Patrono, C. (1997). *In vivo* formation of 8-epi-prostaglandin F2 alpha is increased in hypercholesterolemia. *Arterioscler. Thromb. Vasc. Biol.* **17,** 3230–3235.

Davi, G., Guagnano, M. T., Ciabattoni, G., Basili, S., Falco, A., Marinopiccoli, M., Nutini, M., Sensi, S., and Patrono, C. (2002). Platelet activation in obese women: Role of inflammation and oxidant stress. *JAMA* **288,** 2008–2014.

Davies, S. S., Zackert, W., Luo, Y., Cunningham, C. C., Frisard, M., and Roberts, L. J., 2nd. (2006). Quantification of dinor, dihydro metabolites of F2-isoprostanes in urine by liquid chromatography/tandem mass spectrometry. *Anal. Biochem.* **348,** 185–191.

Dietrich, M., Block, G., Hudes, M., Morrow, J. D., Norkus, E. P., Traber, M. G., Cross, C. E., and Packer, L. (2002). Antioxidant supplementation decreases lipid peroxidation biomarker F(2)-isoprostanes in plasma of smokers. *Cancer Epidemiol. Biomarkers Prev.* **11,** 7–13.

Famm, S. S., and Morrow, J. D. (2003). The isoprostanes: Unique products of arachidonic acid oxidation—a review. *Curr. Med. Chem.* **10,** 1723–1740.

Gniwotta, C., Morrow, J. D., Roberts, L. J., 2nd, and Kuhn, H. (1997). Prostaglandin F2-like compounds, F2-isoprostanes, are present in increased amounts in human atherosclerotic lesions. *Arterioscler. Thromb. Vasc. Biol.* **17,** 3236–3241.

Gopaul, N. K., Anggard, E. E., Mallet, A. I., Betteridge, D. J., Wolff, S. P., and Nourooz-Zadeh, J. (1995). Plasma 8-epi-PGF2 alpha levels are elevated in individuals with non-insulin dependent diabetes mellitus. *FEBS Lett.* **368,** 225–229.

Gross, M., Steffes, M., Jacobs, D. R., Jr., Yu, X., Lewis, L., Lewis, C. E., and Loria, C. M. (2005). Plasma F2-isoprostanes and coronary artery calcification: The CARDIA Study. *Clin. Chem.* **51,** 125–131.

Halliwell, B., and Gutteridge, J. M. (1990). Role of free radicals and catalytic metal ions in human disease: An overview. *Methods Enzymol.* **186,** 1–85.

Haschke, M., Zhang, Y. L., Kahle, C., Klawitter, J., Korecka, M., Shaw, L. M., and Christians, U. (2007). HPLC-atmospheric pressure chemical ionization MS/MS for quantification of 15-F2t-isoprostane in human urine and plasma. *Clin. Chem.* **53,** 489–497.

Kadiiska, M. B., Gladen, B. C., Baird, D. D., Germolec, D., Graham, L. B., Parker, C. E., Nyska, A., Wachsman, J. T., Ames, B. N., Basu, S., Brot, N., Fitzgerald, G. A., *et al.*

(2005). Biomarkers of oxidative stress study II: Are oxidation products of lipids, proteins, and DNA markers of CCl4 poisoning? *Free Radic. Biol. Med.* **38,** 698–710.

Keaney, J. F., Jr., Larson, M. G., Vasan, R. S., Wilson, P. W., Lipinska, I., Corey, D., Massaro, J. M., Sutherland, P., Vita, J. A., and Benjamin, E. J. (2003). Obesity and systemic oxidative stress: Clinical correlates of oxidative stress in the Framingham Study. *Arterioscler. Throm. Vasc. Biol.* **23,** 434–439.

Liang, Y., Wei, P., Duke, R. W., Reaven, P. D., Harman, S. M., Cutler, R. G., and Heward, C. B. (2003). Quantification of 8-iso-prostaglandin-F(2alpha) and 2,3-dino-8-iso-prostaglandin-F(2alpha) in human urine using liquid chromatography-tandem mass spectrometry. *Free Radic Biol. Med.* **34,** 409–418.

Masoodi, M., and Nicolaou, A. (2006). Lipidomic analysis of twenty-seven prostanoids and isoprostanes by liquid chromatography/electrospray tandem mass spectrometry. *Rapid Commun. Mass Spectrom.* **20,** 3023–3029.

Milne, G. L., Musiek, E. S., and Morrow, J. D. (2005). F2-isoprostanes as markers of oxidative stress *in vivo*: An overview. *Biomarkers* **10**(Suppl. 1), S10–S23.

Milne, G. L., Sanchez, S. C., Musiek, E. S., and Morrow, J. D. (2007). Quantification of F_2-isoprostanes as a biomarker of oxidative stress. *Nat. Protocols* **2,** 221–226.

Montine, K. S., Quinn, J. F., Zhang, J., Fessel, J. P., Roberts, L. J., 2nd, Morrow, J. D., and Montine, T. J. (2004). Isoprostanes and related products of lipid peroxidation in neurodegenerative diseases. *Chem. Phys. Lipids* **128,** 117–124.

Montine, T. J., Beal, M. F., Robertson, D., Cudkowicz, M. E., Biaggioni, I., O'Donnell, H., Zackert, W. E., Roberts, L. J., and Morrow, J. D. (1999). Cerebrospinal fluid F2-isoprostanes are elevated in Huntington's disease. *Neurology* **52,** 1104–1105.

Montine, T. J., Markesbery, W. R., Morrow, J. D., and Roberts, L. J., 2nd (1998). Cerebrospinal fluid F2-isoprostane levels are increased in Alzheimer's disease. *Ann. Neurol.* **44,** 410–413.

Morrow, J. D. (2005). Quantification of isoprostanes as indices of oxidant stress and the risk of atherosclerosis in humans. *Arterioscler. Throm. Vasc. Biol.* **25,** 279–286.

Morrow, J. D., Frei, B., Longmire, A. W., Gaziano, J. M., Lynch, S. M., Shyr, Y., Strauss, W. E., Oates, J. A., and Roberts, L. J., 2nd. (1995). Increase in circulating products of lipid peroxidation (F2-isoprostanes) in smokers. Smoking as a cause of oxidative damage. *N. Engl. J. Med.* **332,** 1198–1203.

Morrow, J. D., Harris, T. M., and Roberts, L. J., 2nd (1990a). Noncyclooxygenase oxidative formation of a series of novel prostaglandins: Analytical ramifications for measurement of eicosanoids. *Anal. Biochem.* **184,** 1–10.

Morrow, J. D., Hill, K. E., Burk, R. F., Nammour, T. M., Badr, K. F., and Roberts, L. J., 2nd (1990b). A series of prostaglandin F2-like compounds are produced *in vivo* in humans by a non-cyclooxygenase, free radical-catalyzed mechanism. *Proc. Natl. Acad. Sci. USA* **87,** 9383–9387.

Morrow, J. D., Chen, Y., Brame, C. J., Yang, J., Sanchez, S. C., Xu, J., Zackert, W. E., Awad, J. A., and Roberts, L. J. (1999a). The isoprostanes: Unique prostaglandin like products of free radical-catalyzed lipid peroxidation. *Drug Metab. Rev.* **31,** 117–139.

Morrow, J. D., Zackert, W. E., Yang, J. P., Kurhts, E. H., Callewaert, D., Dworski, R., Kanai, K., Taber, D., Moore, K., Oates, J. A., and Roberts, L. J. (1999b). Quantification of the major urinary metabolite of 15-F2t-isoprostane (8-iso-PGF2alpha) by a stable isotope dilution mass spectrometric assay. *Anal. Biochem.* **269,** 326–331.

Morrow, J. D., and Roberts, L. J. (1999). Mass spectrometric quantification of F_2-isoprostanes in biological fluids and tissues as a measure of oxidant stress. *Methods Enzymol.* **300,** 3–12.

Pratico, D., Barry, O. P., Lawson, J. A., Adiyaman, M., Hwang, S. W., Khanapure, S. P., Iuliano, L., Rokach, J., and FitzGerald, G. A. (1998). IPF2alpha-I: An index of lipid peroxidation in humans. *Proc. Natl. Acad. Sci. USA* **95,** 3449–3454.

Roberts, L. J., 2nd, Moore, K. P., Zackert, W. E., Oates, J. A., and Morrow, J. D. (1996). Identification of the major urinary metabolite of the F2-isoprostane 8-iso-prostaglandin F2alpha in humans. *J. Biol. Chem.* **271,** 20617–20620.

Rokach, J., Khanapure, S. P., Hwang, S. W., Adiyaman, M., Lawson, J. A., and FitzGerald, G. A. (1997a). The isoprostanes: A perspective. *Prostaglandins* **54,** 823–851.

Rokach, J., Khanapure, S. P., Hwang, S. W., Adiyaman, M., Lawson, J. A., and FitzGerald, G. A. (1997b). Nomenclature of isoprostanes: A proposal. *Prostaglandins* **54,** 853–873.

Sircar, D., and Subbaiah, P. V. (2007). Isoprostane measurement in plasma and urine by liquid chromatography-mass spectrometry with one-step sample preparation. *Clin. Chem.* **53,** 251–258.

Taylor, A. W., Bruno, R. S., Frei, B., and Traber, M. G. (2006). Benefits of prolonged gradient separation for high-performance liquid chromatography-tandem mass spectrometry quantitation of plasma total 15-series F-isoprostanes. *Anal. Biochem.* **350,** 41–51.

Wang, Z., Ciabattoni, G., Creminon, C., Lawson, J., Fitzgerald, G. A., Patrono, C., and Maclouf, J. (1995). Immunological characterization of urinary 8-epi-prostaglandin F2 alpha excretion in man. *J. Pharmacol. Exp. Ther.* **275,** 94–100.

Zhang, B., and Saku, K. (2007). Control of matrix effects in the analysis of urinary F$_2$-isoprostanes using novel multidimensional solid-phase extraction and LC-MS/MS. *J. Lipid Res.* **18,** 733–744.

CHAPTER SEVEN

MEASUREMENT OF PRODUCTS OF DOCOSAHEXAENOIC ACID PEROXIDATION, NEUROPROSTANES, AND NEUROFURANS

Kyle O. Arneson *and* L. Jackson Roberts, II

Contents

Abstract

Free radicals derived primarily from oxygen have been implicated in the pathophysiology of a wide variety of human diseases. Quantification of products of free radical damage in biological systems is necessary to understand the role of free radicals in disease states. Measures of lipid peroxidation are often used to quantitate oxidative damage though many of these measures have inherent problems with sensitivity and specificity especially when used to quantitate *in vivo* oxidative injury. The discovery of the F_2-isoprostanes (F_2-IsoPs), prostaglandin F_2-like compounds derived by the free radical peroxidation of

Division of Clinical Pharmacology, Vanderbilt University School of Medicine, Nashville, Tennessee

Methods in Enzymology, Volume 433
ISSN 0076-6879, DOI: 10.1016/S0076-6879(07)33007-3

arachidonic acid (AA, C20:4, ω-6) has largely overcome these limitations. The measurement of the F_2-IsoPs has been shown to be one of the most accurate approaches to quantifying oxidative damage *in vivo*. We have extended our studies of lipid peroxidation and the F_2-IsoPs to docosahexaenoic acid (DHA, C22:6, ω-3) and its peroxidation products. We have found that DHA oxidizes both *in vitro* and *in vivo* to form F_2-IsoP-like compounds termed F_4-neuroprostanes (F_4-NPs). DHA is specifically enriched in neuronal membranes making the F_4-NPs sensitive and specific markers of neuronal oxidative damage. Adapting the methodology used to quantitate the F_2-IsoPs, we utilize stable isotope dilution, negative ion chemical ionization, gas chromatography mass spectrometry (GC/MS) to quantitate the F_4-NPs with a limit of detection in the low picomolar range. Methods have been developed to quantitate both the F_4-NPs and the neurofurans (NFs), DHA derived peroxidation products containing a substituted tetrahydrofuran ring, in brain tissue and cerebrospinal fluid. This review outlines in detail proper sample handling, extraction and hydrolysis of the F_4-NPs and NFs from tissue membrane phospholipids or biological fluids, and purification and derivatization of the compounds for analysis by GC/MS.

1. INTRODUCTION

Free radicals derived primarily from oxygen have been implicated in the pathophysiology of a wide variety of human disease including atherosclerosis, neurodegenerative diseases, cancer, and even the normal aging process (Halliwell and Gutteridge, 1990). Initially, much of the evidence for the role of free radicals and oxidative damage in these disease processes was indirect or circumstantial, mainly because of the limitations in the methodology available to directly quantitate free radical production and their reaction products in biological systems. This particular problem is made more prominent when noninvasive approaches are used to assess oxidative damage in animals or humans (Halliwell and Grootveld, 1987).

Measures of lipid peroxidation are one measure of oxidative damage frequently utilized to implicate free radicals in pathophysiology. These measurements include quantification of short chain alkanes, malondialdehyde, or conjugated dienes. Each of these, however, suffers from problems related to specificity and sensitivity, especially when utilized to quantitate oxidative damage *in vivo*. Part of the difficultly is that many of these lipid peroxidation products have rapid and robust metabolism and innate chemical instability and reactivity. Further, artifactual generation of these lipid peroxidation products can occur *ex vivo* and factors such as endogenous metabolism can affect levels of compounds measured (Halliwell and Grootveld, 1987).

Many of these limitations in measuring oxidative damage and lipid peroxidation were addressed with the discovery of prostaglandin (PG) F_2-like compounds, termed F_2-isoprostanes (F_2-IsoPs) by our lab in 1990

(Morrow *et al.*, 1990). F$_2$-IsoPs (Fig. 7.1A) are produced *in vivo* in humans by a noncyclooxygenase free radical–catalyzed mechanism involving the peroxidation of the polyunsaturated fatty acid (PUFA) arachidonic acid (AA, C20:4, ω-6). Formation of these compounds initially involves the generation of four positional peroxyl radical isomers of arachidonate which undergo endocyclization to PGG$_2$-like compounds. These intermediates are reduced to form four F$_2$-IsoP regioisomers, each which can consist of eight racemic diastereomers. The PGG$_2$-like intermediate is also able to undergo isomerization to form D$_2$/E$_2$-IsoPs (Morrow *et al.*, 1994) which dehydrate to form J$_2$/A$_2$-IsoPs (Chen *et al.*, 1999) and rearrange to form thromboxane-like compounds (Morrow *et al.*, 1996). The ratio of F-ring to D/E-ring compounds is determined by the reducing environment in which the IsoPs form (Morrow *et al.*, 1998). A greater reducing environment favors the formation of the F$_2$-IsoPs over that of the D$_2$/E$_2$-IsoPs. These studies have shown that oxidation of PUFAs results in an extremely complex array of products, including those that resemble the enzymatically produced eicosanoids that can be isolated, characterized, and quantitated.

The measurement of the F$_2$-IsoPs has been shown to be one of the most accurate approaches to measuring oxidative injury *in vivo* (Kadiiska *et al.*, 2005; Roberts and Morrow, 2000). The F$_2$-IsoPs are chemically and metabolically stable and nonreactive, and the artifactual formation of F$_2$-IsoPs *ex vivo* by autoxidation can be prevented by rapid storage of samples at −70° or below and the use of antioxidants during the F$_2$-IsoP assay. Therefore, measurement of F$_2$-IsoPs overcomes the limitations associated with other measures of lipid peroxidation. We utilize gas chromatography/mass spectrometry (GC/MS) to quantitate the F$_2$-IsoPs, the methodology which was described previously in this series (Morrow and Roberts, 1999). More specifically, after isolation and derivatization of the F$_2$-IsoPs, we take advantage of stable isotope dilution, negative ion chemical ionization (NICI) GC/MS with selected ion monitoring (SIM) for

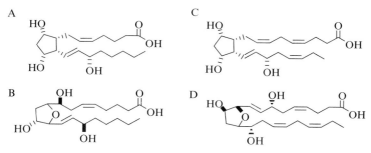

Figure 7.1 Representative structures of AA and DHA lipid peroxidation products. (A) F$_2$-IsoP (15-F$_{2t}$-IsoP). (B) IsoF (*ent*-8-epi-AT-Δ^{13}-9-IsoF). (C) F$_4$-NPs (17-F$_{4c}$-NP). (D) NF (7-epi-AC-Δ^8-10-NF).

quantification. This methodology allows the lower limit of detection of the F_2-IsoPs to be in the low picomolar range. These properties, along with the assay's high sensitivity and specificity, allow the F_2-IsoPs to be an excellent biomarker of and the most robust and sensitive measure of *in vivo* oxidative damage. This notion has been recently independently confirmed in the Biomarkers of Oxidative Stress Study (BOSS) conducted by the National Institutes of Environmental Health Sciences (Kadiiska *et al.*, 2000, 2005a,b).

More recently, we have demonstrated an oxygen insertion step that diverts intermediates from the IsoP pathway to instead form compounds, termed isofurans (IsoFs; Fig. 7.1B) that contain a substituted tetrahydrofuran ring (Fessel *et al.*, 2002). Because of this differential method of formation, we have found that oxygen tensions can affect lipid peroxidation profiles. The formation of IsoFs becomes increasingly favored over that of the IsoPs when oxygen levels are increased. Accordingly we have found that measurement of IsoFs is a much more robust indicator of hyperoxia–induced lung injury than measurement of F_2-IsoPs. Like the IsoPs, the IsoFs are chemically and metabolically stable so are well suited to act as *in vivo* biomarkers of oxidative damage. For example, Parkinson disease has been linked with mitochondrial complex I dysfunction (Schapira, 1998) which can increase relative oxygen tensions within an affected cell. We have shown that the levels of IsoFs are significantly increased in the substantia nigra of these patients while IsoPs are not (Fessel *et al.*, 2003), making the measurement of IsoFs a more sensitive and specific biomarker in this disease state. The ratio of IsoFs to F_2-IsoP also provides information about the relative oxygen tension where the lipid peroxidation is occurring.

The study of lipid peroxidation and the F_2-IsoPs has allowed us to extend our research and methods to other PUFAs and their lipid peroxidation products. Docosahexaenoic acid (DHA, C22:6, ω-3) is an omega-3 PUFA that has been of particular interest. Unlike AA, which is distributed essentially uniformly throughout the body, DHA is preferentially enriched in certain tissues those being the brain, retina, and testes. In the brain, DHA is particularly enriched in the cellular membranes of neurons where it accounts for 25% to 35% of total fatty acids in grey matter (Skinner *et al.*, 1993). The reason that the body would preferentially enrich certain tissues with DHA is not fully understood. DHA-enriched membranes do have differing biophysical properties such as increased fluidity as compared to AA-enriched membranes and it has been shown that receptor function and cellular signaling can be affected by DHA content in cellular membranes (Stillwell and Wassall, 2003). As the human body can only minimally synthesize DHA from essential precursor fatty acids, DHA is obtained mainly through dietary means. Along with eicosapentaenoic acid (EPA, C20:5, ω-3), DHA is a main constituent of fish oil. DHA deficiency has been linked to slowed mental development (Connor *et al.*, 1992) while

DHA supplementation has been linked to a variety of health benefits including increased cardiovascular health (Mozaffarian and Rimm, 2006) and decreased rates of neurodegenerative diseases (Schaefer *et al.*, 2006). Because of the physiologic effects, the neuronal enrichment, and the link between neurodegeneration and lipid peroxidation, the study of DHA and its oxidation products have been important in the understanding of both the normal physiology and the pathophysiology of the brain.

We have reported that DHA can undergo oxidation both *in vitro* and *in vivo* resulting in the formation of a series of F_2-IsoP-like molecules termed F_4-neuroprostanes (F_4-NPs; Roberts *et al.*, 1998). The F_4-NPs (Fig. 7.1C) are very similar in structure to the F_2-IsoPs, their differences deriving from the differences in the precursor parent fatty acids. DHA, besides being an omega-3 fatty acid, has two additional carbon–carbon double bonds and its carbon chain is two carbons longer than that of AA, and these differences are seen in the structure of the F_4-NPs. The increased units of unsaturation present in DHA allows for there to be more possible sites of free radical attack, meaning DHA is more susceptible to lipid peroxidation than AA. Thus, when equal amounts of DHA and AA are oxidized *in vitro*, relatively more F_4-NPs are formed. The F_4-NPs consist of eight regioisomers whose formation is dependent on the initial site of free radical attack and each regioisomer consists of eight racemic diastereomers, for a total of 128 possible compounds. Besides the F_4-NPs, the oxidation of DHA has been shown to result in the formation of other IsoP-like compounds including D_4/E_4-NPs (Reich *et al.*, 2000) and J_4/A_4-NPs (Fam *et al.*, 2002).

The study of NPs allows a unique insight into oxidative damage in the central nervous system (CNS). This is particularly important because of the implication of oxidative damage and lipid peroxidation being causative factors in numerous neurodegenerative diseases (Montine *et al.*, 2004). In the basal state, glial cells out number neurons 10 to 1, and this ratio is further increased in neurodegenerative diseases as neurons die and glia proliferate and hypertrophy. Therefore, markers of oxidative damage measured in brain tissue obtained from patients with neurodegenerative disease will be mainly assessing oxidative damage that has occurred in the glia. The NPs are the only quantitative *in vivo* biomarker of oxidative damage that is selective for neurons.

Recent studies have focused on identifying and characterizing the DHA derived, IsoF-like compounds, termed neurofurans (NFs). We have found that the NFs (Fig. 7.1D) are indeed formed from the peroxidation of DHA and that they are able to be purified and quantitated. Our initial studies show that their formation and use as a biomarker of oxidative damage is slightly more complicated than that of the F_4-NPs and the IsoFs. This is because DHA is more unsaturated than AA, but more importantly that it is an omega-3 fatty acid. The reason for this increased complexity is under active investigation. One important characteristic of the NFs is that they co-purify with the F_4-NPs using the methodology that will be described here. Therefore, quantification of the NFs requires no additional sample preparation to be quantified along with the F_4-NPs.

In this chapter, we will outline the methodology to analyze biological samples focusing on brain tissue and cerebrospinal fluid (CSF) for F_4-NPs utilizing mass spectrometric techniques. Procedures to quantitate both free F_4-NPs and F_4-NPs esterified to phospholipids will be covered. F_4-NPs in tissues are formed *in situ* (Morrow *et al.*, 1992), esterified to phospholipids before being hydrolyzed to their free form. Since the F_4-NPs must be in their free form to be quantitated by GC/MS, first the tissue phospholipids must be extracted and subjected to base hydrolysis to release free F_4-NPs. Finally, specific examples and special considerations relating to the F_4-NPs and the NFs will be provided.

2. Handling and Storage of Biological Fluids and Tissues for Quantification of F_4-Neuroprostanes

Because the formation of the F_4-NPs and other lipid peroxidation products is a nonenzymatic process, care must be taken in order to prevent lipid peroxidation products from forming artifactually during the handling and storage of the biological samples. F_4-NPs and other lipid peroxidation products can be generated *ex vivo* in biological samples in which DHA-containing lipids are present. This occurs not only when biological fluids are left at room temperature, but also when the samples are stored at $-20°$ (Morrow and Roberts, 1997), because tissue fluid is not solid ice at $-20°$. There are several steps that one can take to prevent this extraneous lipid peroxidation from occurring. We have found that the *ex vivo* formation of F_4-NPs is minimal if the biological fluids or tissues are frozen immediately after procurement and if agents including butylated hydroxytoluene (BHT, a free radical scavenger) and/or triphenylphospine (a reducing agent) are added to the organic solvents during extraction and hydrolysis of phospholipids.

3. Extraction and Hydrolysis of F_4-Neuroprostane-Containing Phospholipids in Tissues and Biological Fluids

As mentioned earlier, the formation of the F_4-NPs occurs *in situ* in the phospholipid bilayer and then subsequently released in free form. This creates two populations of F_4-NPs, F_4-NPs that remain esterified in membranes and free F_4-NPs that have been hydrolyzed and released in free form.

To quantitate total F_4-NP formation, both free and esterified F_4-NPs are analyzed. It is necessary to extract the phospholipids from the tissue or biological fluid and release the F_4-NPs from the phospholipid backbone via base hydrolysis (Fig. 7.2).

Five ml of ice-cold Folch solution ($CHCl_3$/methanol, 2:1, v/v), containing 0.005% BHT in a 17- × 100 mm polypropylene culture tube, are added to 0.01 to 1 g of tissue. As discussed above, the presence of BHT during extraction and hydrolysis of lipids is important since it inhibits *ex vivo* formation of F_4-NPs during this portion of the assay. A second 5-ml aliquot of ice-cold Folch solution is added to a separate culture tube. The tissue is then homogenized with a blade homogenizer (Polytron PT 1200E; Kinematica AG, Switzerland) for approximately 30 s. The second aliquot of ice-cold Folch solution is used to wash the blade homogenizer and to ensure that all sample tissue is recovered as tissue can adhere to or become lodged inside the blade of the homogenizer. The two aliquots are then combined and the mixture is allowed to stand sealed under nitrogen at room temperature for 1 h, vortexing occasionally.

Two milliliters of aqueous NaCl (0.9%) are then added and the solution vigorously vortexed before being centrifuged at $800g$ for 5 min. After centrifugation, the upper aqueous layer is discarded and the lower organic layer carefully separated from the intermediate semisolid proteinaceous layer. The organic layer and the proteinaceous layer can be readily separated

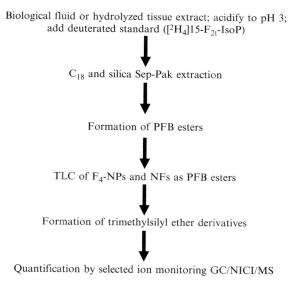

Biological fluid or hydrolyzed tissue extract; acidify to pH 3;
add deuterated standard ($[^2H_4]15$-F_{2t}-IsoP)

↓

C_{18} and silica Sep-Pak extraction

↓

Formation of PFB esters

↓

TLC of F_4-NPs and NFs as PFB esters

↓

Formation of trimethylsilyl ether derivatives

↓

Quantification by selected ion monitoring GC/NICI/MS

Figure 7.2   Outline of the procedure used for the extraction, purification, derivatization, and mass spectrometric analysis of F_4-NPs and NFs from biological sources.

by carefully pouring off the organic layer into a 50-ml conical centrifuge tube, leaving the proteinaceous layer behind in the culture tube. This method works especially well when the proteinaceous layer is large. When the proteinaceous layer is small because of the type and size of the tissue sample, it is often easier to remove the aqueous and proteinaceous layers simultaneously via suction. Care must be taken, however, not to compromise the organic phase if this approach is used.

After the organic phase has been transferred to a 50-ml conical centrifuge tube, it is evaporated to dryness under a stream of nitrogen. One milliliter of methanol containing BHT (0.005%) and 1 ml of aqueous KOH (15%) are then added to the residue. The samples are incubated at 37° for 30 min to effect hydrolysis and release of the F_4-NPs. The mixture is then acidified to pH 3 with 1 M HCl and diluted to a final volume of 20 ml with pH 3 water in preparation for purification and derivatization of free F_4-NPs as described in the next section. Dilution of the methanol in the solution with water to 5% or less is necessary to ensure proper column extraction of F_4-NPs in the subsequent purification procedure. Now, the F_4-NPs are present in their free form, which is necessary for analysis by GC/MS.

For the extraction of lipids containing esterified F_4-NPs in biological fluids such as CSF or plasma, as opposed to tissue, a slightly different method is used. We have found that the addition of BHT alone to the Folch solution used to extract lipids from plasma does not suppress *ex vivo* lipid peroxidation entirely. To fully prevent autoxidation of the lipids during this portion of the assay, the addition of both triphenylphosphine and BHT is necessary to the organic solvents used to extract the lipids. Thus, 20 ml of ice-cold Folch solution ($CHCl_3$/methanol 2:1, v/v) containing 0.005% BHT and triphenylphosphine (50 mg/100 ml Folch solution) are added to 1 ml of a biological fluid in a 50-ml conical centrifuge tube. The mixture is shaken or vortexed for 1 to 2 min, and then 10 ml of 0.043% $MgCl_2$ in water are added. The mixture is vortexed well for another minute and then centrifuged at 800g for 5 min. The upper aqueous phase is discarded and the lower organic phase is separated from the proteinaceous layer as described above. The organic phase is poured into a 50-ml conical centrifuge tube, dried under nitrogen, and the residue is subjected to base hydrolysis using the same procedure outlined above for tissue lipids.

Though in the past we quantitated total (free + bound) F_4-NPs in biological fluids such as CSF and plasma, we have found that the correlation between free and total F_4-NPs is very linear. This linear relationship makes the measurement of total F_4-NPs in biological fluids no more informative than measuring free F_4-NPs only. Since measuring the free compounds is more time and resource efficient, we rarely quantitate esterified F_4-NPs in biological fluids anymore, instead preferring the measurement of free compounds only. The assay to measure free F_4-NPs is detailed in the next section.

4. PURIFICATION, DERIVATIZATION, AND QUANTIFICATION OF F_4-NEUROPROSTANES

Quantification of F_4-NPs by gas chromatography/negative-ion chemical ionization (NICI) mass spectrometry is extremely sensitive with a lower limit of detection in the range of ~5 pg using a deuterated internal standard with a $^2H_0$ blank of less than 2 parts per thousand. Thus it is usually not necessary to assay more than 1 to 2 ml of a biological fluid such as CSF or a lipid extract from more than 50 to 100 mg of tissue. The following assay procedure is adaptable to both hydrolyzed lipid extracts from tissues as described in the previous section where total F_4-NPs are measured, and for biological fluids where free F_4-NPs only are measured. The main difference is in the starting volume of the samples. Normally, 1 to 2 ml of CSF are processed for analysis, whereas 20 ml of the hydrolyzed tissue extracts are processed as described in the previous section.

Following acidification of the sample to pH 3 with 1 M HCl, 200 to 1000 pg of deuterated standard is added. The internal standard is a deuterium-labeled isoprostane, $[^2H_4]15$-F_{2t}-IsoP. This standard is available commercially ($[^2H_4]$-8-iso-prostaglandin F_2; Cayman Chemical, Ann Arbor, MI). The amount of internal standard added depends on the levels of F_4-NPs in the sample as well as the sensitivity of the mass spectrometer. For low-level samples such as CSF, less internal standard needs to be added. Samples that consist of a particularly large amount of tissue will require more internal standard. This is because complex tissues such as brain, despite our best purification efforts, will still contain some unwanted compounds that may potentially have the same m/z ratio as the internal standard when analyzed by GC/MS. Increasing the amount of internal standard to 1000 or even 2000 pg in these samples minimizes the variability in the internal standard ion channel because of contamination in the tissue sample.

The complexity of brain tissue also presents an additional unique concern when larger samples of brain are processed. Because of the high lipid content of brain tissue, it is possible for the solid-phase extraction columns to become clogged, possibly to the point where the sample would have to be abandoned. To prevent this, we have added an additional centrifugation step when working with larger brain samples, which has been found to greatly reduce the problem of clogged columns. After the internal standard is added to the 20 ml of acidified hydrolyzed brain tissue extract, the sample is centrifuged at 800g for 2 to 3 min. After centrifugation, a white, lipid-rich film will be placed on top of the sample and is subsequently aspirated off. In addition, a small pellet will form at the bottom of the 50-ml conical centrifuge tube that will remain in the tube after the sample is applied to the column. The exclusion of the upper white film and the lower pellet is sufficient to prevent clogging of the extraction columns in most instances.

This additional centrifugation step is always performed after the addition of the internal standard to ensure that if there is any loss of F_4-NPs in this step, the loss will be paralleled by a similar loss of internal standard, and thus the ratio of F_4-NPs to internal standard will not change. We have not noticed any difference in F_4-NP levels or peak patterns in samples where this extra centrifugation step has been used.

After addition of the internal standard, the mixture is vortexed and applied slowly to a C_{18} Sep-Pak cartridge (Waters Associates, Milford, MA) preconditioned with 5 ml of methanol and 7 ml of water (pH 3). A 10-ml plastic syringe is used to elute the sample and subsequent solvents through the Sep-Pak cartridge. Once the sample has been added, the column is washed sequentially with 10 ml of water (pH 3) and 10 ml of heptane, which removes nonpolar contaminates including unoxidized DHA. The F_4-NPs are eluted with 10 ml of ethyl acetate/heptane (50:50, v/v) into a 20-ml scintillation vial. The ethyl acetate/heptane eluate from the C_{18} Sep-Pak is then dried over anhydrous Na_2SO_4 and applied to a silica Sep-Pak cartridge (Waters Associates), which has been preconditioned with 5 ml of ethyl acetate. This step should be completed promptly, as Na_2SO_4 has been shown to adsorb lipids to some degree. Care is taken not to transfer any Na_2SO_4 to the silica Sep-Pak cartridge. The cartridge is washed with ethyl acetate/heptane (75:25, v/v) and the F_4-NPs are eluted with 5 ml of ethyl acetate/methanol (50:50, v/v) into a 5-ml glass reacti-vial (Pierce, Rockford, IL). The ethyl acetate/methanol eluate is evaporated under a stream of nitrogen to dryness.

The F_4-NPs are then converted to pentafluorobenzyl (PFB) esters by treatment with a mixture of 40 μl of 10% pentafluorobenzyl bromide in acetonitrile and 20 μl of 10% N,N-diisopropylethylamine in acetonitrile at 37° for 20 min. The reagents are then dried under nitrogen and the residue dissolved in 50 μl methanol/$CHCl_3$ (3:2, v/v) and subjected to thin-layer chromatography (TLC) using a solvent system of chloroform/ethanol (93:7, v/v) and silica gel 60A plates (LK6D, Whatman, Florham Park, NJ). The TLC is run to approximately 13 cm. Approximately 5 μg of the methyl ester of $PGF_{2\alpha}$ is chromatographed on a separate lane and visualized by spraying with a 10% solution of phosphomolybdic acid in ethanol followed by heating. Compounds migrating in the region of the methyl ester of $PGF_{2\alpha}$ ($R_f \sim 0.15$) and the areas 1 cm below and 4 cm above are scraped and extracted from the silica gel with 1 ml of ethyl acetate. The required TLC cut is much wider than that taken for the F_2-IsoPs.

The ethyl acetate is dried under nitrogen and the F_4-NPs are then converted to trimethylsilyl ether derivatives by adding 20 μl N,O-bis (trimethylsilyl)trifluoroacetamide (BSTFA) and 8 μl of dimethylformamide and incubating at 37° for 5 min. The reagents are dried under nitrogen and the F_4-NPs are redissolved in $\sim$20 μl of undecane, which has been dried over calcium hydride, for analysis by GC/MS. Again, like the amount of internal standard added to a sample, the amount of undecane used to dissolve the derivatized sample depends on the levels of F_4-NPs. Samples rich in F_4-NPs will require additional undecane to keep the sample from overloading the column during GC. Likewise, low-level samples require less undecane in order for the GC/MS signal to be of sufficient intensity to quantitate.

For analysis of F_4-NPs, we routinely use a Hewlett Packard 5982A GC/MS system interfaced with an IBM Pentium computer. GC is performed using a 15-m, 0.25-mm diameter, 0.25-μm film thickness, DB1701 fused silica capillary column (Fisons, Folsum, CA). The column temperature is programmed from 190 to 290° at 20°/min. Methane is used as the carrier gas for NICI at a flow rate of 1 ml/min. Ion source temperature is 250°, electron energy is 70 eV, and the filament current is 0.25 mA.

The major ion generated in the NICI mass spectra of the pentafluorobenzyl ester, tris-trimethylsilyl ether derivatives of F_4-NP is m/z 593, which represents the M-181 (M-$CH_2C_6F_5$) carboxylate anion. The corresponding M-181 ion for the $[^2H_4]15$-F_{2t}-IsoP internal standard is m/z 573. The ionization efficiency of the parent compounds to their M-181 ion has been found to be greater than 95%. Quantification of the F_4-NP levels is achieved through selected ion monitoring (SIM) and comparing the area of the appropriate peaks in the m/z 593 SIM chromatogram of the F_4-NPs to that of the single peak of the internal standard in the m/z 573 SIM chromatogram.

This methodology for F_4-NPs and other lipid peroxidation products such as isoprostanes is not as time efficient as other methods of analysis such as immunoassays. The increased time is, however, more than compensated for and required to take advantage of the increased sensitivity and specificity of the assay. Approximately 12 to 20 samples can be assayed for F_4-NP per day using these methods by a trained individual. At times though, it may be helpful to complete the assay over the course of 2 days and there are several points where the assay can be suspended. The sample prep could be suspended after elution off either of the Sep-Paks or after the sample is in ethyl acetate before silylation. The samples should be sealed and stored at −20°.

4.1. Analysis of biological samples containing low levels of F_4-neuroprostanes

It is common for biological samples, especially those derived from the brain, to be limited in size and contain very low levels of F_4-NPs. Therefore, we have optimized our F_4-NP assay to still be able to reliably quantitate very low levels by ensuring maximum recovery of F_4-NPs from the tissue. First, less internal standard, 100 to 500 pg, is added to the sample when F_4-NP levels are predicted to be low to allow the standard to be at a comparable concentration as the F_4-NPs in the sample. Next, to ensure that all of the

F_4-NPs are recovered, the sample is eluted off the silica Sep-Pak with 10 ml of ethyl acetate/methanol (50:50, v/v). In addition, after the TLC plates have been scraped, the silica is extracted twice with ethyl acetate and the two extractions combined. Finally, the samples are dissolved in a smaller volume of undecane, 6 to 10 μl to increase the concentration of F_4-NPs of the sample that will be injected into the GC/MS. Special low-volume glass inserts (National Scientific, Rockwood, TN) are available for mass spectrometry autosampling vials. These steps as well as optimizing the sensitivity of the mass spectrometer have allowed us to easily quantitate F_4-NPs in as little as 1 ml of CSF and $\sim$5 mg of brain tissue. If the sample is too concentrated for efficient analysis by the GC/MS, additional undecane can be added directly to the sample to lower the F_4-NP concentration into the linear range of the instrument.

4.2. Application of the F_4-neuroprostane assay in the analysis of CNS tissue and fluids

As mentioned previously, we have successfully used this assay to quantify F_4-NPs in brain tissue as well as CSF in both animal and human samples. Shown in Fig. 7.3 is a selected ion current chromatogram (SIM) obtained from the analysis of F_4-NPs generated from oxidation of DHA *in vitro*. The series of peaks in the middle m/z 593 SIM chromatogram represent the different endogenous F_4-NPs. This pattern of peaks is representative to that obtained in biological tissues or fluids that we have analyzed. In the lower m/z 573 SIM chromatogram, the single peak represents the $[^2H_4]15\text{-}F_{2t}$-IsoP internal standard that was added to the sample. For quantification purposes, the F_4-NPs peaks denoted by the horizontal line are routinely measured. Using the ratio of the integrated F_4-NP peak areas denoted, to the area of the internal standard peak, the concentration of F_4-NPs is calculated.

4.3. Simultaneous quantification of F_2-IsoPs and F_4-NPs

Often we analyze our biological samples for lipid peroxidation products derived from both AA and DHA simultaneously, especially when dealing with brain tissue or CSF, as this provides insight into the source and location of the oxidative damage. As mentioned earlier, DHA is highly enriched in neuronal membranes, while AA is found in both neuronal and glial cells. Therefore, the measurement of F_4-NPs provides a more selective index of neuronal oxidative damage, while the measurement of F_2-IsoPs provides an index of global oxidative damage occurring in the brain, integrating data from both glial and neuronal cells.

It is not always possible to obtain two identical samples or samples in sufficient quantity to be able to analyze separate samples for F_4-NPs

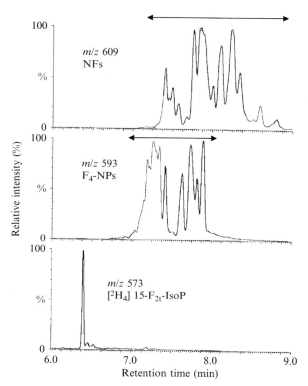

Figure 7.3 Analysis of oxidized DHA for F$_4$-NPs and NF. DHA was oxidized *in vitro* utilizing an iron/ADP/ascorbate oxidation system. The lower *m/z* 573 ion–current chromatogram represents the [^{2}H$_4$] 15-F$_{2t}$-IsoP internal standard. The middle *m/z* 593 ion–current chromatogram represents the F$_4$-NPs. The peaks in the middle chromatogram under the *horizontal arrow* are integrated for quantification of the F$_4$-NPs. The upper *m/z* 609 ion–current chromatogram represent the NFs. The peaks under the *horizontal arrow* are integrated to quantitate the NFs.

and F$_2$-IsoPs. The similarity between the two assays and the use of a common internal standard allows us to use a single sample to measure both families of compounds by splitting the sample before the TLC step. To do this the sample is processed as described for the normal F$_4$-NP assay, but the sample is dissolved in twice the amount of CH$_3$OH/CHCl$_3$ (3:2, v/v) (100 μl) after conversion to PFB esters. Then 50 μl are spotted on two separate lanes and the TLC plates are scraped accordingly, 1 cm above and below the methyl ester PGF$_{2\alpha}$ standard for the F$_2$-IsoPs and 1 cm below and 4 cm above the standard for the F$_4$-NPs. This effectively splits a biological sample into two, allowing for individual analysis of both F$_2$-IsoPs and F$_4$-NPs.

4.4. Quantification of biological samples for neurofurans

As mentioned earlier, the NFs have only recently been identified and characterized so it is not yet clear if they will have unique advantages over that of the F_4-NPs as biomarkers of oxidative damage. The methodology presented above also allows for the quantification of the NFs as they co-purify with the F_4-NPs. The NFs are also quantitated as pentafluoro-benzyl ester, tris-trimethylsilyl ether derivatives with m/z of 609 represent-ing the M-181 ion. The upper tracing in Fig. 7.3, the m/z 609 SIM chromatogram is that of the NFs. NF levels are quantitated by comparing the area of the peaks in SIM m/z 609 to the area of the internal standard peak in m/z 573. A single sample can therefore be quantitated by GC/MS simultaneously for both F_4-NPs and the NFs by monitoring both m/z 593 and 609 ions. Thus, when analyzing for F_4-NPs no additional work is required to analyze a sample for NFs.

4.5. Optimization of the F_4-neuroprostane assay

Although our methods to quantitate the F_4-NPs are highly sensitive and specific, the methodology described above is not a fully optimized assay. The described assay is a modification of the assay we employ to quantitate the F_2-IsoPs. While this methodology produces very reproducible results, the assay variability for F_4-NPs has been seen to be slightly higher than that of the F_2-IsoPs. The reasons for the higher variability can be at least partially attributed to use of a F_2-IsoP, $[^2H_4]15$-F_{2t}-IsoP, as the internal standard to quantify F_4-NPs. While this IsoP is structurally similar to the F_4-NPs, the former is slightly more polar and therefore will have slightly different extraction and chromatographic characteristics than the F_4-NPs. This may reduce the accuracy of this method to quantify NPs from sample to sample. Also, the quantification of the F_4-NPs requires the integration of the peak areas of multiple compounds eluting from the GC over a 2- to 3-min interval. This increases the possibility for compounds that are not F_4-NPs to co-elute and interfere with F_4-NPs quantification. The likelihood of co-eluting compounds is increased in the F_4-NP assay because of its wider TLC cut. Finally the fact that the quantification of the F_4-NPs requires the integration of multiple peaks representing only a fraction of the regio- and stereo-isomer formed can decrease the accuracy and precision of this assay. Small differences in the preparations of the TLC solvents or characteristics of different silica gel TLC plates may also introduce variability in the assay. In spite of these caveats, measurement of F_4-NPs by GC/MS represents one of the most reliable and accurate methods to assess oxidative neuronal injury in the brain.

Members of our group have reported an assay for F_4-NPs that has been further optimized using synthesized 17-F_{4C}-NP in which two $[^{18}O_2]$ atoms have been incorporated into the carboxyl group for use as an internal standard (Musiek *et al.*, 2004). The TLC solvent system has also been adjusted to chloroform/ethanol (90:10, v/v, run up to 15 cm), and the methyl ester of $17\text{-}F_{4C}\text{-NP}$ is used as the TLC standard. This allows a narrower TLC cut to be taken, which is from immediately below the TLC standard to 2 cm above. Quantification is achieved by comparing the area of the peaks in the m/z 593 SIM chromatogram of the F_4-NP to the area of the single peak in the m/z 597 SIM chromatogram of the $^{18}O_2$-labeled F_4-NP standard. This optimized assay was shown to reduce the variability in the quantification of the F_4-NPs. As the $^{18}O_2$-labeled $17\text{-}F_{4C}\text{-NP}$ standard is not commercially available, the original F_4-NP assay is still currently the one of choice. This new method has laid the groundwork to transition in the future to a F_4-NP assay that is fully optimized.

5. SUMMARY

This chapter has outlined methods to assess lipid peroxidation associated with oxidative injury occurring *in vivo* within the CNS by quantifying concentrations of free F_4-NPs in CSF and levels of esterified F_4-NPs esterified in lipids in brain tissue. The mass spectrometric assay employed is highly sensitive and specific.

By extending our knowledge of the F_2-IsoPs to the lipid peroxidation products of DHA, the F_4-NPs have provided a unique insight into the oxidative damage occurring within the CNS. Few if any biomarkers of oxidative damage can claim the neuronal sensitivity and specificity of the F_4-NPs especially in the setting of neurodegenerative diseases. With the limitations inherent in animal models and postmortem studies, analysis of F_4-NP levels in human CSF provides a powerful approach to advance our understanding of the role of oxidative damage in diseases of the CNS.

ACKNOWLEDGMENTS

Supported by National Institutes of Health grants GM42056 (MERIT Award to LJR) and 5 T32 GM07347 (Medical Scientist Training Program Grant). K. O. A. was a recipient of a PhRMA Foundation Medical Student Research Fellowship.

REFERENCES

Chen, Y., Morrow, J. D., and Roberts, L. J., 2nd (1999). Formation of reactive cyclopentenone compounds *in vivo* as products of the isoprostane pathway. *J. Biol. Chem.* **274,** 10863–10868.
Connor, W. E., Neuringer, M., and Reisbick, S. (1992). Essential fatty acids: The importance of n-3 fatty acids in the retina and brain. *Nutr. Rev.* **50,** 21–29.

Fam, S. S., Murphey, L. J., Terry, E. S., Zackert, W. E., Chen, Y., Gao, L., Pandalai, S., Milne, G. L., Roberts, L. J., Porter, N. A., Montine, T. J., and Morrow, J. D. (2002). Formation of highly reactive A-ring and J-ring isoprostane-like compounds (A4/J4-neuroprostanes) *in vivo* from docosahexaenoic acid. *J. Biol. Chem.* **277,** 36076–36084.

Fessel, J. P., Hulette, C., Powell, S., Roberts, L. J., 2nd, and Zhang, J. (2003). Isofurans, but not F2-isoprostanes, are increased in the substantia nigra of patients with Parkinson's disease and with dementia with Lewy body disease. *J. Neurochem.* **85,** 645–650.

Fessel, J. P., Porter, N. A., Moore, K. P., Sheller, J. R., and Roberts, L. J., 2nd (2002). Discovery of lipid peroxidation products formed *in vivo* with a substituted tetrahydrofuran ring (isofurans) that are favored by increased oxygen tension. *Proc. Natl. Acad. Sci. USA* **99,** 16713–16718.

Halliwell, B., and Grootveld, M. (1987). The measurement of free radical reactions in humans. Some thoughts for future experimentation. *FEBS Lett.* **213,** 9–14.

Halliwell, B., and Gutteridge, J. M. (1990). Role of free radicals and catalytic metal ions in human disease: An overview. *Methods Enzymol.* **186,** 1–85.

Kadiiska, M. B., Gladen, B. C., Baird, D. D., Dikalova, A. E., Sohal, R. S., Hatch, G. E., Jones, D. P., Mason, R. P., and Barrett, J. C. (2000). Biomarkers of oxidative stress study: Are plasma antioxidants markers of CCl(4) poisoning? *Free Radic. Biol. Med.* **28,** 838–845.

Kadiiska, M. B., Gladen, B. C., Baird, D. D., Germolec, D., Graham, L. B., Parker, C. E., Nyska, A., Wachsman, J. T., Ames, B. N., Basu, S., Brot, N., Fitzgerald, G. A., *et al.* (2005a). Biomarkers of oxidative stress study II: Are oxidation products of lipids, proteins, and DNA markers of CCl4 poisoning? *Free Radic. Biol. Med.* **38,** 698–710.

Kadiiska, M. B., Gladen, B. C., Baird, D. D., Graham, L. B., Parker, C. E., Ames, B. N., Basu, S., Fitzgerald, G. A., Lawson, J. A., Marnett, L. J., Morrow, J. D., Murray, D. M., *et al.* (2005b). Biomarkers of oxidative stress study III. Effects of the nonsteroidal anti-inflammatory agents indomethacin and meclofenamic acid on measurements of oxidative products of lipids in CCl4 poisoning. *Free Radic. Biol. Med.* **38,** 711–718.

Montine, K. S., Quinn, J. F., Zhang, J., Fessel, J. P., Roberts, L. J., 2nd, Morrow, J. D., and Montine, T. J. (2004). Isoprostanes and related products of lipid peroxidation in neurodegenerative diseases. *Chem. Phys. Lipids* **128,** 117–124.

Morrow, J. D., Awad, J. A., Boss, H. J., Blair, I. A., and Roberts, L. J., 2nd (1992). Non-cyclooxygenase-derived prostanoids (F2-isoprostanes) are formed *in situ* on phospholipids. *Proc. Natl. Acad. Sci. USA* **89,** 10721–10725.

Morrow, J. D., Awad, J. A., Wu, A., Zackert, W. E., Daniel, V. C., and Roberts, L. J., 2nd (1996). Nonenzymatic free radical-catalyzed generation of thromboxane-like compounds (isothromboxanes) *in vivo*. *J. Biol. Chem.* **271,** 23185–23190.

Morrow, J. D., Hill, K. E., Burk, R. F., Nammour, T. M., Badr, K. F., and Roberts, L. J., 2nd (1990). A series of prostaglandin F2-like compounds are produced *in vivo* in humans by a non-cyclooxygenase, free radical-catalyzed mechanism. *Proc. Natl. Acad. Sci. USA* **87,** 9383–9387.

Morrow, J. D., Minton, T. A., Mukundan, C. R., Campbell, M. D., Zackert, W. E., Daniel, V. C., Badr, K. F., Blair, I. A., and Roberts, L. J., 2nd (1994). Free radical-induced generation of isoprostanes *in vivo*. Evidence for the formation of D-ring and E-ring isoprostanes. *J. Biol. Chem.* **269,** 4317–4326.

Morrow, J. D., and Roberts, L. J. (1997). The isoprostanes: Unique bioactive products of lipid peroxidation. *Prog. Lipid Res.* **36,** 1–21.

Morrow, J. D., and Roberts, L. J., 2nd (1999). Mass spectrometric quantification of F2-isoprostanes in biological fluids and tissues as measure of oxidant stress. *Methods Enzymol.* **300,** 3–12.

Morrow, J. D., Roberts, L. J., Daniel, V. C., Awad, J. A., Mirochnitchenko, O., Swift, L. L., and Burk, R. F. (1998). Comparison of formation of D2/E2-isoprostanes and F2-isoprostanes *in vitro* and *in vivo*—effects of oxygen tension and glutathione. *Arch. Biochem. Biophys.* **353,** 160–171.

Mozaffarian, D., and Rimm, E. B. (2006). Fish intake, contaminants, and human health: Evaluating the risks and the benefits. *JAMA* **296,** 1885–1899.

Musiek, E. S., Cha, J. K., Yin, H., Zackert, W. E., Terry, E. S., Porter, N. A., Montine, T. J., and Morrow, J. D. (2004). Quantification of F-ring isoprostane-like compounds (F4-neuroprostanes) derived from docosahexaenoic acid *in vivo* in humans by a stable isotope dilution mass spectrometric assay. *J. Chromatogr. B Analyt. Technol. Biomed. Life Sci.* **799,** 95–102.

Reich, E. E., Zackert, W. E., Brame, C. J., Chen, Y., Roberts, L. J., 2nd, Hachey, D. L., Montine, T. J., and Morrow, J. D. (2000). Formation of novel D-ring and E-ring isoprostane-like compounds (D4/E4-neuroprostanes) *in vivo* from docosahexaenoic acid. *Biochemistry* **39,** 2376–2383.

Roberts, L. J., 2nd, Montine, T. J., Markesbery, W. R., Tapper, A. R., Hardy, P., Chemtob, S., Dettbarn, W. D., and Morrow, J. D. (1998). Formation of isoprostane-like compounds (neuroprostanes) *in vivo* from docosahexaenoic acid. *J. Biol. Chem.* **273,** 13605–13612.

Roberts, L. J., and Morrow, J. D. (2000). Measurement of F(2)-isoprostanes as an index of oxidative stress *in vivo. Free Radic. Biol. Med.* **28,** 505–513.

Schaefer, E. J., Bongard, V., Beiser, A. S., Lamon-Fava, S., Robins, S. J., Au, R., Tucker, K. L., Kyle, D. J., Wilson, P. W., and Wolf, P. A. (2006). Plasma phosphatidyl-choline docosahexaenoic acid content and risk of dementia and Alzheimer disease: The Framingham Heart Study. *Arch. Neurol.* **63,** 1545–1550.

Schapira, A. H. (1998). Mitochondrial dysfunction in neurodegenerative disorders. *Biochim. Biophys. Acta* **1366,** 225–233.

Skinner, E. R., Watt, C., Besson, J. A., and Best, P. V. (1993). Differences in the fatty acid composition of the grey and white matter of different regions of the brains of patients with Alzheimer's disease and control subjects. *Brain* **116,** 717–725.

Stillwell, W., and Wassall, S. R. (2003). Docosahexaenoic acid: Membrane properties of a unique fatty acid. *Chem. Phys. Lipids* **126,** 1–27.

CHAPTER EIGHT

Enantiomeric Separation of Hydroxy and Hydroperoxy Eicosanoids by Chiral Column Chromatography

Claus Schneider, Zheyong Yu, William E. Boeglin, Yuxiang Zheng, *and* Alan R. Brash

Contents

Abstract

We describe high-performance liquid chromatography (HPLC) methods for the enantiomeric resolution of hydroxy and hydroperoxy fatty acids/eicosanoids using a Chiralpak AD or AD-RH chiral stationary phase. These columns achieve baseline resolution of all six positional/conjugated isomers of the hydroxy as well as of the hydroperoxy derivatives of arachidonic acid in chromatographic runs of less than 20 min. Hydro(pero)xy derivatives of linoleic and linolenic

Division of Clinical Pharmacology, Department of Pharmacology, Vanderbilt University School of Medicine, Nashville, Tennessee

Methods in Enzymology, Volume 433

ISSN 0076-6879, DOI: 10.1016/S0076-6879(07)33008-5

acids can be resolved with similar efficiencies. The individual hydroperoxy isomers are best resolved using the reversed-phase Chiralpak AD-RH column. For the synthesis of milligram quantities of enantiomerically pure hydro(pero)xy arachidonic acids, a simple scheme is presented starting with the autoxidation of the fatty acid methyl ester in the presence of 10% α-tocopherol followed by chromatographic purification of the positional isomers using a combination of reversed- and straight-phase HPLC columns. Mild alkaline hydrolysis of the methyl ester derivatives affords the free acids suitable for biological testing. The Chiralpak AD column appears to be efficient for the chiral resolution of prostaglandins and isoprostanes although a comprehensive evaluation is yet to be reported. For chiral analysis of endogenous hydroxy eicosanoids the availability of novel microflow Chiralpak capillary columns (o.3 mm i.d.) will be of great advantage, because sample sizes of a few nanograms can be analyzed using simple UV detection.

1. INTRODUCTION

Hydroperoxy and hydroxy eicosanoids can be formed via enzymatic pathways in animals and plants, but also through nonenzymatic lipid peroxidation (Brash, 1999; Niki *et al.*, 2005; Roberts and Morrow, 2002; Rouzer and Marnett, 2003; Thoma *et al.*, 2004). Enzymatic oxygenation of fatty acids is usually associated with high stereospecificity of the oxygenation resulting in a chirally pure product, but exceptions to this concept have been observed. For example, formation of the 15*RS*-hydroxyeicosatetraenoic acid (15*RS*-HETE) by-product during prostaglandin synthesis by the cyclooxygenase enzymes proceeds with low level of stereocontrol, and ratios of 70:30 up to racemic mixtures have been observed, depending on the source and the cyclooxygenase isozymes studied (Schneider *et al.*, 2002; Xiao *et al.*, 1997). On the other hand, autoxidative processes lead to formation of racemic mixtures of products, especially of the initially formed hydroperoxides, but one should keep in mind that this balance may be shifted upon preferential metabolism of one enantiomer over the other. A documented example is the faster oxidation of the 4*R*-enantiomer of the cytotoxic lipid peroxidation product, 4-hydroxynonenal, in brain mitochondria (Honzatko *et al.*, 2005). Nevertheless, chiral analysis generally provides a good measure for the enzymatic origin of metabolites of interest, and we will present simple HPLC methods for chiral analysis of hydroxy eicosanoids that can be performed with standard equipment available in the analytical laboratory.

The history of chiral column technology shows several striking advances over the past twenty years of application to analysis of eicosanoids and other hydroxy fatty acids. First came the Pirkle phase columns which achieved, at best, baseline resolution of enantiomers, although most commonly poorly resolved double peaks (Kühn *et al.*, 1987). Advent of the Chiralcel phases produced some significant improvements, although initially with broad chromatographic peaks and lack of a useful separation for preparative purposes (Brash and Hawkins, 1990). The most recent advance to be demonstrated with this class of compounds is the versatile Chiralpak AD columns, normal phase or reversed phase (RP), and which form the main feature of this article. We should note, however, that there is plenty of room for improvement for some of the separations discussed here, and the ongoing development of chiral phase (CP) technology (currently promoted as the Chiralpak 1A and 1B phases, http://www.chiraltech.com/new/faq. htm) is a continuing process to be expected and anticipated for the future. The development of novel microflow columns is another facet of the technological advances, and one which offers greatly improved sensitivity for analytical application (http://www.chiraltech.com/new/faq.htm.)

 ## 2. Chiral Columns, Solvents, and Standards

2.1. Selection of a chiral column

Current chiral column technology is capable of excellent chiral discrimination for most hydroxy eicosanoids. The problem is identifying the appropriate column for the task at hand. This would be expensive to accomplish by trial and error, given that the most versatile columns sell for about three or four times the cost of a first-rate C18 or silica column. If sufficient sample is available in pure form, it is worth making use of the technical support facilities at Chiral Technologies; the company will screen their battery of columns for the best match (www.chiraltech.com/new/faq.htm). The company's most versatile and successful CPs include the Chiralcel OD and the Chiralpak AD, the latter including separate normal-phase AD and RP AD-RH columns. There is a good chance that at least one will achieve chiral resolution, although the selection process remains a gamble.

The AD-column chiral stationary phase is based on a *tris*(3,5–dimethyl-phenyl carbamate)-derivatized amylose support that is coated on the silica base. The designation "H" signifies smaller particle–sized material and higher chromatographic efficiency (sharper peaks). The new generation of Chiralpak 1A (same chiral polymer as Chiralpak AD) and Chiralpak 1B (i.e., Chiralcel OD) columns uses an immobilized chiral stationary phase, and therefore come with the promise of higher stability to a broader range of solvents with the same or better separation performance.

If only one chiral column is to be selected, the Chiralpak AD-RH may be the best choice. It is operated in the RP mode, can accommodate solvent mixtures of acetonitrile/water and methanol/water, and is also resistant to the use of small amounts of acid in the mobile phase. Inclusion of an acid modifier in the solvent is necessary if fatty acid derivatives are analyzed without prior methylation or any other derivatization of the carboxylate. The Chiralpak AD, which is run in straight-phase (SP) mode, works well for hydroxy fatty acids, but is generally less successful in resolution of hydroperoxy derivatives.

2.2. Chiralpak AD and effects of the alcohol modifier

In running hexane-based solvent systems, one would normally consider that it makes little difference which alcohol modifier is used to adjust retention times for products of differing polarities. It turns out, however, that for the Chiralcel AD column, separation is critically dependent on the alcohol selected. Although the overall retention volumes are similar, chiral resolution is greatly improved on switching from isopropanol (poor resolution) to ethanol (intermediate) to methanol (the best). This effect was first reported for chiral separations of several classes of aromatics and pharmaceuticals (e.g., Kunath *et al.*, 1996; Tang, 1996). Subsequently, we established its importance for resolution of various hydroxy fatty acids (Schneider *et al.*, 2000), and the effect has been confirmed with more polar eicosanoids such as the prostaglandins and isoprostanes (Lee *et al.*, 2003; Yin *et al.*, 2003, 2007). In essence, feeble or weak separations using isopropanol (IPA) are transformed to excellent by substitution of ethanol or, even better, methanol. Methanol is miscible with hexane at 5% v/v, but is biphasic at 10% methanol in hexane. In our experience, the hexane/methanol 100:5 (v/v) mixtures can also become biphasic on standing in the cooler temperatures in the laboratory overnight, so caution is warranted. Methanol/IPA or ethanol/IPA mixtures as modifier have been used to elute more polar eicosanoids (Lee *et al.*, 2003; Yin *et al.*, 2007). A theoretical basis for the effect of alcohol modifier has been proposed, essentially a hydrogen bonding interaction if a prominent alcohol-dependent effect is observed (Zhao and Pritts, 2007). The mechanism underlying the enantiomeric discrimination in the AD chiral phase has been studied for certain aromatic alcohols and a rational model proposed (Yamamoto *et al.*, 2002), although in general the rational prediction of chiral separations remains elusive.

To achieve reproducible separations, it is important that the Chiralpak AD column is equilibrated in the mobile phase for 30 min or longer. In our experience, it takes a long equilibration time to convert back from a hexane/IPA-based solvent to a solvent with ethanol or methanol as the alcohol modifier. Small variations in solvent composition can have a marked influence on retention times, and we have seen large increases in retention, especially using hexane-based mixtures.

For the RP AD column, we have not noticed any difference in chiral resolution in using methanol or ethanol in the alcohol/water solvents. Because alcohol/water mixtures are viscous, and given that the AD-RH packing is pressure sensitive, it may be preferable to use methanol to help avoid pressure problems. As the percent water goes over 10% and the solvent becomes more viscous, flow rate on standard-sized (0.46 cm i.d.) columns should be cut to 0.5 ml/min or less.

2.3. Determination of the elution order

A simple way of determining the elution order is to compare retention times with chirally pure standards. Standards are available for common hydro (pero)xy eicosanoids from commercial sources like Cayman Chemical, Biomol, and Larodan. In addition, having the racemic mixture at hand is helpful, too, if one can refer to the elution order from published work. When we initially determined the elution order for H(P)ETEs, H(P)ODEs, and H(P)OTEs, we used chiral products of known configuration formed in enzymatic reactions using lipoxygenase and cyclooxygenase enzymes. Alternatively, CD spectroscopy of suitable derivatives can be used to determine the absolute configuration of hydroxy compounds, and it provides an independent confirmation of the chromatographic results. Finally, we found that the R enantiomer eluted before the S enantiomer for almost all hydroxy and hydroperoxy products we analyzed using the Chiralpak AD and AD-RH columns. The R before S elution order is maintained in both normal-phase and RP-AD columns.

2.4. Preparative synthesis of enantiomeric hydroperoxyeicosatetraenoic acid isomers

Only a small selection of fatty acid hydroperoxides are available commercially (e.g., hydroperoxyeicosatetraenoic acids [HPETEs], hydroperoxyctadecadienoic acids [HODEs]), partly because of lack of demand for the products and partly because only a few can be produced using readily available enzymes (15S-HPETE, 12S-HPETE). The following method for production of equimolar mixtures of racemic fatty acid hydroperoxides is one of the easiest, most reliable in analytical biochemistry (Peers and Coxon, 1983). It yields a clean mixture of hydroperoxy fatty acid methyl esters through autoxidation (and subsequently these can be hydrolyzed to the corresponding hydroperoxy free acids). The required starting materials are the pure fatty acid methyl ester (e.g., arachidonate methyl ester, from Sigma or Nu-Chek Prep Inc., Elysian, MN), and α-tocopherol (Sigma), also known as vitamin E. (Although the method can be applied to the corresponding free fatty acids, in this case the hydroperoxide products are partly—about one-third—reduced to the hydroxy derivatives, presenting a complex HPLC profile unless the whole sample is reduced before chromatography.)

α-Tocopherol (Sigma) is sold as a dark reddish oil, about 67% by weight as the active isomer. It can be quantified by weight, or using its distinctive absorbance in the UV (λ_{max} at 292 nm, 0.76 AU = 100 μg/ml). Although known mainly as an antioxidant, in this method the α-tocopherol functions not to prevent the initial oxygenation of the fatty acid, but to stabilize the resulting peroxyl radicals as hydroperoxides, and thus block subsequent rearrangement reactions of the primary autoxidation products. As a consequence, these otherwise unstable fatty acid hydroperoxides accumulate during the 3- to 5-day course of the autoxidation. The method is suitable for production of any level of product, depending only on the amount of fatty acid starting material. But after a point the subsequent purifications become difficult (with tailing and cross-contamination of peaks on HPLC). For preparation of small amounts of reference materials (say, 0.1 mg of each fatty acid hydroperoxide isomer), it is much easier to start with modest levels of substrate (e.g., 1 to 10 mg).

Arachidonic acid methyl ester, 0.1 mg to 1 g, is dissolved in a small volume of volatile solvent, transferred to a round-bottom flask and mixed with 10% (w/w) of α-tocopherol. The solvent is evaporated under a stream of nitrogen, the flask filled with oxygen, capped, and placed in an oven at 37°. The oxygen is replenished daily, at which time the progress of the autoxidation can be followed using UV spectroscopy. After approximately 3 days, the lipid is dissolved in a suitable solvent such as chloroform or methylene chloride, a small aliquot taken, dissolved in ethanol, and scanned from 350 and 200 nm using a UV/Vis spectrophotometer. The hydroperoxide products contain a conjugated diene and display a chromophore with λmax around 235 nm (0.75AU $\sim$ 10 μg/ml). Based on our experience, $\sim$30% of product can be accumulated before further transformation reactions of the hydroperoxides lead to a loss in product yield (these transformation reactions are evident by an increase of the absorbance at 270 nm). The racemic hydroperoxides obtained, six from arachidonate (5, 8, 9, 11, 12, and 15) or two from linoleate (9 and 13) are readily resolved by SP-HPLC (Peers and Coxon, 1983; Porter *et al.*, 1979) and partially resolved by RP-HPLC. The 1979 paper by Porter and coworkers (1979) contains a very useful table that lists the SP-HPLC order of elution of HETEs, HPETEs, HETE methyl esters, and HPETE methyl esters.

Conversion of the methyl esters of hydroperoxy fatty acids to the corresponding free acids is potentially tricky, because a typical alkaline hydrolysis procedure can degrade the hydroperoxide, producing the hydroxy fatty acid and other derivatives (Gardner *et al.*, 1996). To avoid this problem, the exposure time to alkali should be kept to a minimum. To help with reducing the contact time, the hydroperoxide methyl ester concentration should be dilute, thus minimizing micelle formation which will protect part of the sample from the aqueous alkaline phase and effectively prolong the time to completion of hydrolysis. Inclusion of dichloromethane in the reaction mixture also helps considerably in preventing micelles and reducing contact time. Occasional sonication will break up micelles and have a similar effect. Another key to high recoveries is to keep the sample cold (on ice) before reaction and throughout the extraction procedure, and maintain in a nitrogen or argon atmosphere when standing. Hydroperoxy fatty ester (<1 mg/ml) in methanol/dichloromethane (10:1, v/v) is brought to room temperature, an equal volume of 1M KOH is added, mixed and the sample kept at room temperature under nitrogen or argon with occasional sonication in a water bath. After 30 min, the sample is acidified to pH <5 and extracted with an equal volume of dichloromethane. The organic phase is washed twice with water, and then transferred to a dry vial (leaving behind any drops of water in the original vial). The dichloromethane is evaporated to dryness under a stream of nitrogen, and the dried sample redissolved in methanol or ethanol immediately upon evaporation of the last of the solvent. Ideally, the purity of the preparation is checked by SP-HPLC (typical solvent, hexane/IPA/glacial acetic acid, 100:2:0.1, by volume). The most likely contaminant is unhydrolyzed hydroperoxy methyl ester. With practice, overall recoveries of 80 to 90% can be obtained, even after SP-HPLC purification of the hydroperoxide free fatty acid. HPETEs and HETEs are quantified by UV spectroscopy using the conjugated diene chromophore at 235 nm ($\varepsilon = 23,000$ (Gibian and Vandenberg, 1987), meaning that 10 μg/ml $\approx$ 0.75 AU). Fatty acid hydroperoxides should be stored in the alcohol solvent under nitrogen or argon at $-20°$ or below.

3. Chiral Analyses

3.1. Hydroperoxy fatty acids using HPLC with UV detection

Chiral resolution of the HPETE methyl esters is achieved using the Chiralpak AD-RH column (15 × 0.46 cm) eluted with a solvent of methanol/water (88:12 v/v) at a 1-ml/min flow rate. Baseline resolution is achieved for all six conjugated HPETE isomers as shown in Fig. 8.1. In each of the panels about 0.5 mg of HPETE were injected on the column. Semi-preparative separation of the methyl ester of 11-HPETE (~3 mg injected) is shown in Fig. 8.2. In this example, a 25 × 1.0 cm Chiralpak AD column was eluted with a solvent of hexane/methanol (100:2 v/v) at a flow rate of 4 ml/min.

3.2. Hydroxy fatty acids using HPLC with UV detection

Chiral analysis should be preceded by chromatographic isolation of the hydroxy fatty acids of interest. Co-elution of the product of interest with an authentic standard on HPLC is a good control for the correct identification of the product. It also protects the chiral column from exposure to contaminants from the original sample or the methylation reaction. For isolation of the eicosanoid product, we usually employ RP-HPLC as the first step, using methanol/water/acetic acid mixtures of 85/15/0.01 or 80/20/0.01 (v/v) as

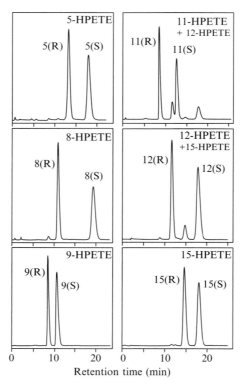

Figure 8.1 Reversed-phase chiral HPLC of racemic HPETE methyl esters. All products were resolved using a Chiralpak AD-RH column (15 × 0.46 cm) with a solvent system of methanol/water (88:12, v/v) run at a flow rate of 1 ml/min. The panels on the *left* show, from the top down, separation of 5*RS*-HPETE, 8*RS*-HPETE, and 9*RS*-HPETE methyl esters, and on the *right*, 11*RS*-HPETE, 12*RS*-HPETE, and 15*RS*-HPETE methyl esters. The minor peaks in some of the HPETE chromatograms are identified on the panels in smaller lettering. Detection: ultraviolet absorbance at 235 nm.

solvents. The peaks of interest are collected, evaporated from the organic solvent, and extracted with methylene chloride. (If more polar products like dihydroxy metabolites or prostaglandins are to be analyzed, extraction on a small C18 cartridge [1 cc, 30 mg, Waters or Varian] and elution with methanol are more efficient.) The extracted samples are dissolved in 20 to 50 μl of methanol in a 1-ml conical glass vial and treated with ethereal diazomethane for 30 s to form the methyl ester derivatives. The methylated products are purified using a silica SP–HPLC column and a solvent of hexane/isopropanol/ acetic acid (100/1/0.1 v/v) is effective for most monohydroxylated isomers; more polar products require up to 10% isopropanol. Table 8.1 provides a summary of the fatty acid derivatives separated, chiral column and solvents used, and approximate retention times.

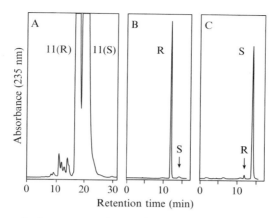

Figure 8.2 Normal-phase preparative resolution of 11*RS*-HPETE methyl ester. (A) Separation of 11*RS*-HPETE methyl ester (3 mg) using a semipreparative Chiralpak AD column (25 × 1.0 cm) eluted with a solvent of hexane/methanol (100:2 v/v) at a flow rate of 4 ml/min. (B) and (C) Chiral purity of the enantiomers collected in (A) was evaluated on small aliquots using an analytical Chiralpak AD column (25 × 0.46 cm) eluted with the same solvent at a flow rate of 1 ml/min. Note that there is only minor crossover of the opposite enantiomer after collection from the semipreparative column.

3.3. Hydroxy fatty acids using HPLC with mass spectrometric detection

Blair and coworkers developed a method for the chiral resolution of hydroxy eicosanoids using the Chiralpak AD column in combination with a mass spectrometric detector (Lee *et al.*, 2003). Conversion of hydroxy metabolites of linoleic and arachidonic acids to their pentafluorobenzyl (PFB) derivatives allowed the use of electron-capture APCI-MS-MS as a highly sensitive and selective detection method. The positional isomers were identified by their characteristic cleavage pattern upon collision-induced dissociation in the selective reaction monitoring mode. The elution of 11-HETE, 12-HETE, 15-HETE, 9-HODE, and 13-HODE, and the prostaglandins PGE_2 and $PGF_{2\alpha}$ in a single run was accomplished by using gradient elution with a mixture of methanol and isopropanol in hexane.

3.4. Saturated hydroxy–fatty acid derivatives

In our limited experience, the separation of saturated fatty acid derivatives is more challenging. Whereas the enantiomers of 9*RS*-HETE or its methyl or PFB esters are readily resolved on AD columns (Lee *et al.*, 2003; Schneider *et al.*, 2000), we found that upon hydrogenation of the double bonds the chromatographic characteristics changed markedly. The enantiomers of 9-HETE PFB ester were resolved using a Chiralcel OD-H column

Table 8.1. Retention time (min)/resolution[a] and elution order of hydroxylated arachidonic and linoleic acids (as methyl ester derivatives) separated on Chiralpak AD column (25 × 0.46 cm) at 1 ml/min using indicated solvent

	Hexane/ethanol (100/2 v/v)			Hexane/methanol (100/2 v/v)		
	(R)	(S)	Resolution	(R)	(S)	Resolution
5-HETE	14.2	15.9	2.2	13.3	18.1	6.9
8-HETE	11.7	13.4	2.8	9.8	12.6	5.2
9-HETE	10.9	11.4	0.8	9.3	9.7	1.0
11-HETE	11.5	13.0	1.6	8.4	10.1	3.3
12-HETE	10.4	11.3	2	8.9	11.3	4.4
15-HETE[c]	9.3	13.6	7.6	10.0	18.7	14.3
	Hexane/ethanol (100/5 v/v)			Hexane/methanol (100/5 v/v)		
	(R)	(S)	Resolution	(R)	(S)	Resolution
9-HODE	6.3	7.4	2.8	6.9	10.9	8.8
13-HODE	5.6	6.5	2.7	5.7	8.6	8.1

[a] Chromatographic resolution, Rs, is defined as the distance between the two peak centers divided by the average of the two peak widths (established as the width at baseline obtained by triangulation). For symmetrical (Gaussian) peaks, when Rs = 1, the cross-contamination between peaks is approximately 2.3%; for Rs = 0.5, cross-contamination is ≈16.5%; and for Rs = 1.5, cross-contamination is ≈0.1%. Reproduced with permission from Schneider *et al.*, 2000.

(retention times 24.7 and 26.9 min using hexane/IPA, 100:2 v/v), whereas the saturated 9*RS*-hydroxyeicosanoate PFB co-eluted as a single peak at 14 min. Similarly, using the AD-RH column, the 9-HETE PFB enantiomers were well resolved (3.7 and 4.6 min using 100% methanol as solvent), whereas the PFB ester of 9-hydroxyeicosanoate gave no separation (single peak at 24.8 min). Separations were better using the Chiralpak AD column (hexane/methanol, 100:2 v/v) with retention times of 12.6 min for 9*R*- and 17.2 min for 9*S*-HETE PFB, whereas under identical chromatographic conditions, the PFB ester of the saturated analog eluted with two barely resolved peaks at 36.5 and 38.7 min.

3.5. Resolution of prostaglandins and isoprostanes

Only a few reports describe use of the Chiralpak AD column for the resolution of prostaglandins or isoprostanes (Yin *et al.*, 2003, 2007). The chiral separations of the pentafluorobenzyl (PFB) ester derivatives of 8,15-di-epi-PGF$_{2\alpha}$ and 12-epi PGF$_{2\alpha}$ were performed as part of a study on the mechanism of isoprostane formation. Racemic PGB$_2$ is very well resolved using the Chiralpak AD column and solvents of hexane/ethanol (90:10, v/v) or hexane/methanol (95:5, v/v) (Schneider *et al.*, 2000). Lee and colleagues (2003) used the Chiralpak AD column on LC-MS to resolve mixtures of eicosanoids as the PFB esters, but chiral separation of the more polar products such as the prostaglandins was not examined. Indeed, a systematic evaluation of the Chiralpak columns for the chiral resolution of prostaglandins has not yet been reported, probably because the racemic standards are not commercially available.

3.6. α,β-Unsaturated 4-hydroxy alkenals

α,β-Unsaturated 4-hydro(pero)xy alkenals are prominent products of the autoxidation of arachidonic and linoleic acids (Uchida, 2003). These alde-hydes are formed by a cleavage of the fatty acid carbon chain. We developed a method for chiral resolution of these aldehydes in order to use the configuration of the hydro(pero)xy group as a tool for studying the mecha-nism of the chain cleavage reaction of fatty acid hydroperoxides (Schneider *et al.*, 2001). The two aldehydes investigated were 4-hydroxy-nonenal and 9-hydroxy-12-oxo-10*E*-dodecenoic acid. The underivatized aldehydes did not elute off the Chiralpak AD column, and therefore were derivatized to the methoxime using a molar excess of O-methoxylamine hydrochloride in pyridine. The resulting *syn* and *anti* isomers were separated using RP-HPLC, and the later eluting *anti* isomer was used for chiral analysis. The *S* and *R* enantiomers of the methoxime derivative of 4-hydroxy-nonenal eluted at 5.7- and 7.4-min retention times, respectively (Chiralpak AD, 25 × 0.46 cm, elution with hexane/ethanol 90:10 at 1.0 ml/min flow rate, UV detection at 235 nm), making this separation the only exception to the observation that the *R* enantiomer always elutes before the *S* enantiomer. The elution order is reversed (*R* before *S*) with the *syn* methoxime deriva-tive of 9-hydroxy-12-oxo-10*E*-dodecenoic acid methyl ester. For both aldehydes, the elution order was established using CD spectroscopy of the 2-naphthoate derivative. When the reversed-phase Chiralpak AD-RH column is used, the free aldehyde can be analyzed directly. However, only a methanol/water solvent resolved a racemic mixture of 4-hydroxy-nonenal (Chiralpak AD-RH, 15 × 0.46 cm, methanol/water 85:15 at 0.5 ml/min). A mixture of acetonitrile and water as the solvent gave a single unresolved peak (Schneider *et al.*, 2004).

4. APPLICATION OF CHIRAL COLUMNS FOR SEPARATION OF DIASTEREOMERS

The availability of a distinctly different chromatographic phase can add to the usual complement of columns available for nonchiral separations. As an example, a challenging separation we encountered recently involved the mixture of the four trihydroxy hydrolysis products of the epoxyalcohol 8R-hydroxy-11R,12R-epoxyeicosa-5Z,10E,14Z-trienoic acid. The four triols almost co-chromatographed on RP-HPLC, and were only partly resolved on SP-HPLC. An efficient separation was achieved by running sequential normal-phase silica and Chiralpak AD columns (Yu *et al.*, 2007).

ACKNOWLEDGMENTS

Supported by National Institutes of Health grants GM-53638, GM-074888, GM-15431, and GM-076592.

REFERENCES

Brash, A. R. (1999). Lipoxygenases: Occurrence, functions, catalysis, and acquisition of substrate. *J. Biol. Chem.* **274,** 23679–23682.

Brash, A. R., and Hawkins, D. J. (1990). High performance liquid chromatography for chiral analysis of eicosanoids. *Methods Enzymol.* **187,** 187–195.

Gardner, H. W., Simpson, T. D., and Hamberg, M. (1996). Mechanism of linoleic acid hydroperoxide reaction with alkali. *Lipids* **31,** 1023–1028.

Gibian, M. J., and Vandenberg, P. (1987). Product yield in oxygenation of linoleate by soybean lipoxygenase: The value of the molar extinction coefficient in the spectrophotometric assay. *Anal. Biochem.* **163,** 343–349.

Honzatko, A., Brichac, J., Murphy, T. C., Reberg, A., Kubatova, A., Smoliakova, I. P., and Picklo, M. J., Sr. (2005). Enantioselective metabolism of trans-4-hydroxy-2-nonenal by brain mitochondria. *Free Radic. Biol. Med.* **39,** 913–924.

Kühn, H., Wiesner, R., Lankin, V. Z., Nekrasov, A., Alder, L., and Schewe, T. (1987). Analysis of the stereochemistry of lipoxygenase-derived hydroxypolyenoic fatty acids by means of chiral phase high-pressure liquid chromatography. *Anal. Biochem.* **160,** 24–34.

Kunath, A., Theil, F., and Jahnisch, K. (1996). Influence of the kind of the alcoholic modifier on chiral separation on a Chiralpak AD column. *J. Chromatogr. A* **728,** 249–257.

Lee, S. H., Williams, M. V., DuBois, R. N., and Blair, I. A. (2003). Targeted lipidomics using electron capture atmospheric pressure chemical ionization mass spectrometry. *Rapid Commun. Mass Spectrom.* **17,** 2168–2176.

Niki, E., Yoshida, Y., Saito, Y., and Noguchi, N. (2005). Lipid peroxidation: Mechanisms, inhibition, and biological effects. *Biochem. Biophys. Res. Commun.* **338,** 668–676.

Peers, K. F., and Coxon, D. T. (1983). Controlled synthesis of monohydroperoxides by a-tocopherol inhibited autoxidation of polyunsaturated lipids. *Chem. Phys. Lipids* **32,** 49–56.

Porter, N. A., Logan, J., and Kontoyiannidou, V. (1979). Preparation and purification of arachidonic-acid hydroperoxides of biological importance. *J. Org. Chem.* **44,** 3177–3181.

Roberts, L. J., 2nd, and Morrow, J. D. (2002). Products of the isoprostane pathway: Unique bioactive compounds and markers of lipid peroxidation. *Cell. Mol. Life Sci.* **59,** 808–820.

Rouzer, C. A., and Marnett, L. J. (2003). Mechanism of free radical oxygenation of polyunsaturated fatty acids by cyclooxygenases. *Chem. Rev.* **103,** 2239–2304.

Schneider, C., Boeglin, W. E., and Brash, A. R. (2000). Enantiomeric separation of hydroxy-eicosanoids by chiral column chromatography: Effect of the alcohol modifier. *Anal. Biochem.* **287,** 186–189.

Schneider, C., Boeglin, W. E., Prusakiewicz, J. J., Rowlinson, S. W., Marnett, L. J., Samel, N., and Brash, A. R. (2002). Control of prostaglandin stereochemistry at the 15-carbon by cyclooxygenases-1 and 2. A critical role for serine 530 and valine 349. *J. Biol. Chem.* **277,** 478–485.

Schneider, C., Porter, N. A., and Brash, A. R. (2004). Autoxidative transformation of chiral ω6 hydroxy linoleic and arachidonic acids to chiral 4-hydroxy-2E-nonenal. *Chem. Res. Toxicol.* **17,** 937–941.

Schneider, C., Tallman, K. A., Porter, N. A., and Brash, A. R. (2001). Two distinct pathways of formation of 4-hydroxynonenal. Mechanisms of nonenzymatic transformation of the 9- and 13-hydroperoxides of linoleic acid to 4-hydroxyalkenals. *J. Biol. Chem.* **276,** 20831–20838.

Tang, Y. B. (1996). Significance of mobile phase composition in enantioseparation of chiral drugs by HPLC on a cellulose-based chiral stationary phase. *Chirality* **8,** 136–142.

Thoma, I., Krischke, M., Loeffler, C., and Mueller, M. J. (2004). The isoprostanoid pathway in plants. *Chem. Phys. Lipids* **128,** 135–148.

Uchida, K. (2003). 4-Hydroxy-2-nonenal: A product and mediator of oxidative stress. *Prog. Lipid Res.* **42,** 318–343.

Xiao, G., Tsai, A. L., Palmer, G., Boyar, W. C., Marshall, P. J., and Kulmacz, R. J. (1997). Analysis of hydroperoxide-induced tyrosyl radicals and lipoxygenase activity in aspirin-treated human prostaglandin H synthase-2. *Biochemistry* **36,** 1836–1845.

Yamamoto, C., Yashima, E., and Okamoto, Y. (2002). Structural analysis of amylose tris (3,5-dimethylphenylcarbamate) by NMR relevant to its chiral recognition mechanism in HPLC. *J. Am. Chem. Soc.* **124,** 12583–12589.

Yin, H., Gao, L., Tai, H. H., Murphey, L. J., Porter, N. A., and Morrow, J. D. (2007). Urinary prostaglandin F2alpha is generated from the isoprostane pathway and not the cyclooxygenase in humans. *J. Biol. Chem.* **282,** 329–336.

Yin, H., Havrilla, C. M., Gao, L., Morrow, J. D., and Porter, N. A. (2003). Mechanisms for the formation of isoprostane endoperoxides from arachidonic acid. "Dioxetane" intermediate versus beta-fragmentation of peroxyl radicals. *J. Biol. Chem.* **278,** 16720–16725.

Yu, Z., Schneider, C., Boeglin, W. E., and Brash, A. R. (2007). Epidermal lipoxygenase products of the hepoxilin pathway selectively activate the nuclear receptor PPARa. *Lipids* **42,** 491–497.

Zhao, Y., and Pritts, W. A. (2007). Chiral separation of selected proline derivatives using a polysaccharide type stationary phase by high-performance liquid chromatography. *J. Chromatogr. A* **1156,** 228–235.

CHAPTER NINE

TARGETED CHIRAL LIPIDOMICS ANALYSIS BY LIQUID CHROMATOGRAPHY ELECTRON CAPTURE ATMOSPHERIC PRESSURE CHEMICAL IONIZATION MASS SPECTROMETRY (LC-ECAPCI/MS)

Seon Hwa Lee *and* Ian A. Blair

Contents

Abstract

The corona discharge used to generate positive and negative ions under conventional atmospheric pressure chemical ionization (APCI) conditions also provides a source of low-energy gas-phase electrons. This is thought to occur by displacement of electrons from the nitrogen sheath gas. Therefore, suitable analytes can undergo electron capture in the gas phase in a manner similar to that observed for gas chromatography/electron capture negative chemical

Centers for Cancer Pharmacology and Excellence in Environmental Toxicology, University of Pennsylvania, Philadelphia, Pennsylvania

Methods in Enzymology, Volume 433
ISSN 0076-6879, DOI: 10.1016/S0076-6879(07)33009-7

ionization/mass spectrometry (MS). This technique, which has been named electron-capture APCI (ECAPCI)/MS, mass spectrometry provides an increase in sensitivity of two orders of magnitude when compared with conventional APCI methodology. It is a simple procedure to tag arachidonic acid– and linoleic acid–derived oxidized lipids with an electron-capturing group such as the pentafluorobenzyl (PFB) moiety before analysis. PFB derivatives have previously been used as electron-capturing derivatives because they undergo dissociative electron capture in the gas phase to generate negative ions through the loss of a PFB radical. A similar process occurs under ECAPCI conditions. By monitoring the negative ions that are formed, it is possible to obtain extremely high sensitivity for PFB derivatives of oxidized lipids derived from arachidonic and linoleic acid. A combination of stable isotope dilution methodology and chiral liquid chromatography-ECAPCI/MS makes it possible to resolve and quantify complex mixtures of regioisomeric and enantiomeric oxidized lipids.

1. INTRODUCTION

Cyclooxygenase (COX)-, cytochrome P450 (CYP)-, and lipoxygenase (LOX)-mediated pathways of arachidonic acid (AA) and linoleic acid (LA) metabolism have been implicated as mediators of numerous diseases including cancer (Jones et al., 2003; Sharma et al., 2001), cardiovascular diseases (Capdevila and Falck, 2002; Wang et al., 2005), and neurodegeneration (Montine and Morrow, 2005; Teismann et al., 2003). These enzymatic pathways result in the formation of oxidized lipids, including prostaglandins (PGs), thromboxanes (TXs), and leukotrienes (LTs), as well as hydroperoxyeicosatetraenoic acids (HPETEs), hydroxyeicosatetraenoic acids (HETEs), hydroperoxyoctadecadienoic acids (HPODEs), hydroxyoctadecadienoic acids (HODEs), and epoxyeicosatrienoic acids (EETs) (Blair, 2001; Capdevila et al., 2002; Chen et al., 2006; Wang and Dubois, 2006). The biological effects of oxidized lipids can occur through direct receptor activation or from homolytic decomposition of HPETEs and HPODEs to bifunctional electrophiles such as 4-oxo-2(E)-nonenal (ONE) (Lee and Blair, 2000) that react with DNA, proteins, and peptides (Blair, 2005, 2006; Lee and Blair, 2001).

Many of the oxidized lipids can also be produced nonenzymatically from the interaction of reactive oxygen species (ROS) with AA and LA esterified into lipids such as phospholipids or with nonesterified AA and LA. In addition, ROS can specifically initiate the formation of isoprostanes (isoPs) from AA esterified in intact glycerolipids and phospholipids (Milne et al., 2005; Rokach et al., 2004). Lipase cleavage of the oxidized lipids from their lipid stores results in the release of their free carboxylate derivatives, which can then potentially induce pathophysiological effects (Musiek *et al.*, 2005). HPETEs, HETEs, HPODEs, HODEs, and EETs are formed as racemic mixtures through nonenzymatic reactions (Porter *et al.*, 1995; Yin and Porter, 2005). Conversely, one enantiomer is predominant when oxidized lipids are generated enzymatically (Williams *et al.*, 2005). Chiral separations can then be used in order to distinguish the nonenzymatic and enzymatic pathways of oxidized lipid formation (Blair, 2005). However, in many cases only trace amounts of individual oxidized lipids are present in the biological fluid that are analyzed and so it is necessary to conduct such determinations with extremely high sensitivity.

2. ELECTRON-CAPTURE ATMOSPHERIC-PRESSURE CHEMICAL IONIZATION

Liquid chromatography (LC) together with atmospheric pressure ionization (API)–based mass spectrometry (MS) methodology has revolutionized our approach to the analysis of biomolecules and drugs (Ackermann *et al.*, 2006; Kamel and Prakash, 2006). LC-atmospheric-pressure chemical ionization (APCI)/MS is a useful API technique because analyte signals are relatively insensitive to suppression by contaminants from the biological matrix (Matuszewski, 2006). Therefore, this technique is often used for accurate and precise analyses of biomolecules, drugs, and their metabolites in biological fluids. Unfortunately, LC-APCI/MS methodology does not have sufficient sensitivity for the trace analysis of oxidized lipids to compete with gas chromatography/electron-capture negative chemical ionization (GC-ECNCI)/MS, which was introduced by Hunt and coworkers in 1976 (Hunt *et al.*, 1976). Therefore, GC-ECNCI/MS, which is a time-consuming and complex technique, has been the method of choice for a number of years for the trace analysis of oxidized lipids derived from AA and LA (Blair, 1990). The pioneering work of Horning and coworkers showed that ionization under APCI conditions was initiated by the $N_2^{+\bullet}$ radical cation, which in turn was formed by collision of high-energy electrons from the corona discharge with the nitrogen sheath gas (Carroll *et al.*, 1975; Horning *et al.*, 1974). However, the potential analytical utility of the low-energy electrons that also result from these collisions was not recognized at the time. We reasoned that the low-energy electrons generated in the APCI source could potentially ionize suitable electron-capturing molecules (such as pentafluorobenzyl [PFB] ester derivatives) through dissociative electron capture (Fig. 9.1). The ionization process would then be analogous to ECNCI in a conventional chemical ionization source (Hunt *et al.*, 1976). The initially formed radical anion would dissociate into a PFB radical and a carboxylate anion (see Fig. 9.1).

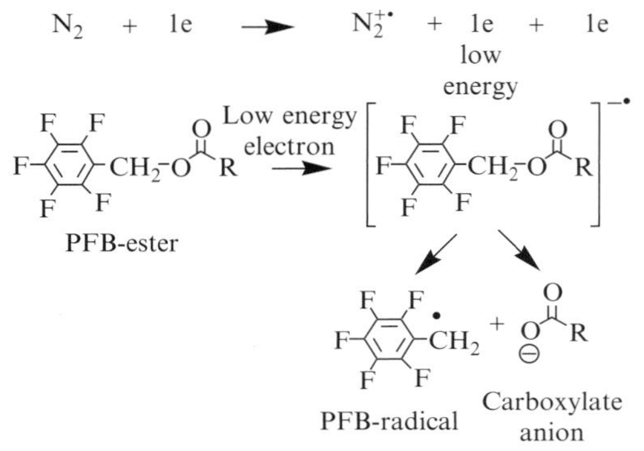

Figure 9.1 Formation of carboxylate negative ions by dissociative electron capture during ECAPCI/MS analysis of PFB-ester derivatives.

We discovered that ECAPCI did indeed occur in the APCI source and that ultrahigh sensitivity could be obtained by analyzing the carboxylate anions derived from PFB ester derivatives of nonesterified oxidized lipids (Singh *et al.*, 2000).

3. DISSOCIATIVE ELECTRON CAPTURE

When 13(*S*)-HODE-PFB was analyzed under negative APCI conditions, an intense negative ion was observed at m/z 295, because of dissociation of a PFB radical from the initially formed radical anion corresponding to [M]$^{-\bullet}$ (Fig. 9.1; Lee *et al.*, 2003; Singh *et al.*, 2000). If conventional APCI had occurred, then a negative ion corresponding to [M-H] would have been observed at m/z 475. In fact, when 13(*S*)-HODE-PFB was analyzed by positive APCI, the expected mass spectrum was obtained in which the protonated molecule (MH$^+$) was observed as a major ion at m/z 477. This meant that under negative APCI conditions, dissociative electron capture had occurred as predicted (Fig. 9.2) and that a novel ionization process had occurred in the source of the mass spectrometer. We named the technique "electron capture (EC) APCI" (Singh *et al.*, 2000). Importantly, the structural integrity of the thermally labile HODE was maintained during the electron capture process. Therefore, we anticipated that this technique would be very useful for highly specific and sensitive analyses as had been demonstrated for the corresponding GC-MS technique of ECNCI (Blair, 1990). Additional analytical specificity was conferred by the use of collision-induced dissociation (CID), which resulted in the formation of product ion (at m/z 195) that was diagnostic for 13-HODE through an α-cleavage had occurred proximal to the 13-hydroxy group (Fig. 9.3). This made it possible to readily distinguish 13 (*S*)-HODE from the other seven HODE regioisomers and stereoisomers that

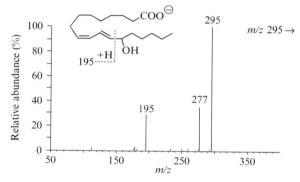

Figure 9.2 ECAPCI/MS analysis of 13(S)-HODE-PFB.

Figure 9.3 Production spectrum after CID of the [M-PFB] negative ion at m/z 295 from 13(S)-HODE.

are formed during lipid peroxidation (Porter *et al.*, 1995). A similar α–cleavage was observed for [²H₄]-13(S)-HODE so that high sensitivity stable isotope dilution quantitative analyses could be conducted using multiple reaction monitoring (MRM) of the transitions m/z 295 → m/z 195 for 13(S)-HODE and m/z 299 → m/z 198 for [²H₄]-13(S)-HODE used as the internal standard. Similar types of specific transitions were observed for many of the other oxidized lipids derived from AA and LA (Lee *et al.*, 2003, 2005a).

4. DERIVATIZATION FOR LC-ECAPCI/MS

In the past, it has proved difficult to conduct a comprehensive analysis of AA- and LA-derived metabolites and oxidation products (oxidized lipids) using GC-MS because of the thermal instability of many of the analytes.

Furthermore, the implementation of LC–MS-based methodology has been hampered by the difficulty in separating isomeric and enantiomeric compounds using conventional reversed-phase chromatography. Thus, most investigators have employed methodology that is based on chiral LC coupled with radioactivity detection of products from radiolabled fatty acid substrates added to disrupted cells or to proteins isolated from cells (Brash *et al.*, 1997; Kamitani *et al.*, 1999). The ability to separate enantiomeric pairs of endogenously generated oxidized lipids with high sensitivity using LC-MS normally represents a formidable challenge. LC-ECAPCI/MS methodology has simplified this task considerably. Most AA- and LA-derived oxidized lipids have a functional group (usually a free acid) that makes it possible to readily attach an electron-capturing derivative. There is a wealth of literature describing highly sensitive GC-ECNCI/MS studies using electron capturing derivatives (Giese, 2000) including, PGs and EETs (Blair, 1990), LTs and HETEs (Strife and Murphy, 1984), isoPs (Morrow *et al.*, 1999), TXs (Barrow *et al.*, 1989), and platelet activating factor (Ramesha and Pickett, 1986). The derivatives used in these studies are all amenable for use with LC-based ECAPCI/MS methodology. We chose the PFB derivative in our LC-ECAPCI/MS-based approach to the analysis of oxidized lipids because it has proved to be extremely robust, and has been used in an enormous number of GC-ECNCI/MS applications (Giese, 2000) since we first described its use for the analysis of endogenous PGs (Blair *et al.*, 1982). It is noteworthy the 2-nitro-4-trifluoromethylphenyl derivative (Higashi *et al.*, 2002) has been used for the analysis of steroids by ECAPCI and that azasteroids and nitroaromatic compounds (Burinsky *et al.*, 2001; Hayen *et al.*, 2002) can be analyzed by ECAPCI without the need for derivatization.

5. Procedure

5.1. Chemicals and materials

9-Oxo-10*E*,12*Z*-octadecadienoic acid (9–oxo–ODE), 13-oxo-ODE, 15-oxo-5*Z*,8*Z*,11*Z*,13*E*-eicosatetraenoic acid (15-oxo-ETE), 9(*R*)-hydroxy-10*E*,12*Z*-octadecadienoic acid (9(*R*)-HODE), 9(*S*)-HODE, 13(*R*)--hydroxy-9*Z*,11*E*-octadecadienoic acid (13(*R*)-HODE), 13(*S*)-HODE, 20-hydroxy-5*Z*,8*Z*,11*Z*,14*Z*-eicosatetraenoic acid (20-HETE), 5(*R*)-hydroxy-6*E*,8*Z*,11*Z*,14*Z*-eicosatetraenoic acid (5(*R*)-HETE), 5(*S*)-HETE, 12(*R*)-hydroxy-5*Z*,8*Z*,10*E*,14*Z*-eicosatetraenoic acid (12(*R*)-HETE), 12(*S*)-HETE, 15(*R*)-hydroxy-5*Z*,8*Z*,11*Z*,13*E*-eicosatetraenoic acid (15(*R*)-HETE), 15 (*S*)-HETE, 11(*R*)-hydroxy-5*Z*,8*Z*,12*E*,14*Z*-eicosatetraenoic acid (11(*R*)-HETE), 11(*S*)-HETE, 8(*R*)-hydroxy-5*Z*,9*E*,11*Z*,14*Z*–eicosatetraenoic acid (8(*R*)-HETE), 8(*S*)-HETE, 5(*S*),12(*R*)-dihydroxy-6*Z*,8*E*,10*E*,

14Z-eicosatetraenoic acid (LTB$_4$), 9-oxo-11α,15S-dihydroxy-prosta-5Z, 13E-dien-1-oic acid (PGE$_2$), 9α,15S-dihydroxy-11-oxo-prosta-5Z, 13E-dien-1-oic acid (PGD$_2$), 9-oxo-11α,15S-dihydroxy-(8β)-prosta-5Z, 13E-dien-1-oic acid (8-iso-PGE$_2$), 9α,11α,15S-trihydroxy-prosta-5Z,13E-dien-1-oic acid (PGF$_{2\alpha}$), 9α,11β,15S-trihydroxy-prosta-5Z,13E-dien-1-oic acid (11β-PGF$_2$), 9α,11α,15S-trihydroxy-(8β)-prosta-5Z,13E-dien-1-oic acid (8-iso-PGF$_{2\alpha}$, iPF$_{2\alpha}$-III), [^{2}H$_4$]-9(S)-HODE, [^{2}H$_4$]-13(S)-HODE, [^{2}H$_6$]-20-HETE, [^{2}H$_8$]-5(S)-HETE, [^{2}H$_8$]-12(S)-HETE, [^{2}H$_8$]-15(S)-HETE, [^{2}H$_6$]-20-HETE, [^{2}H$_4$]-PGE$_2$, [^{2}H$_4$]-PGD$_2$, [^{2}H$_4$]-PGF$_{2\alpha}$, [^{2}H$_4$]-11β-PGF$_2$, and [^{2}H$_4$]-8-iso-PGF$_{2\alpha}$ were purchased from Cayman Chemical Co. (Ann Arbor, MI). Diisopropylethylamine (DIPE) and 2,3,4,5,6-pentafluoro-benzyl bromide (PFB-Br) are purchased from Sigma–Aldrich (St. Louis, MO). Roswell Park Memorial Institute (RPMI) cell culture media and fetal bovine serum (FBS) are supplied by GIBCO (Gland Island, NY). HPLC-grade hexane, methanol, and isopropanol are obtained from Fisher Scientific Co. (Fair Lawn, NJ). Gases are supplied by BOC Gases (Lebanon, NJ).

5.2. Extraction of oxidized lipids from cell culture media

A mixture of 12 heavy isotope internal standards—[^{2}H$_4$]-9(S)-HODE, [^{2}H$_4$]-13(S)-HODE, [^{2}H$_8$]-5(S)-HETE, [^{2}H$_8$]-12(S)-HETE, [^{2}H$_8$]-15(S)-HETE, [^{2}H$_4$]-PGE$_2$, [^{2}H$_4$]-PGD$_2$, and [^{2}H$_4$]-LTB$_4$ (1 ng each), and [^{2}H$_6$]-20-HETE, [^{2}H$_4$]-PGF$_{2\alpha}$, [^{2}H$_4$]-11β-PGF$_2$, and [^{2}H$_4$]-8-iso-PGF$_{2\alpha}$ (10 ng each)—is added to the cell culture media (3 ml). After standing for 10 min to allow for equilibration, the media is acidified to pH 3 with 2.5 N hydrochloric acid, and then extracted with diethyl ether (4 ml × 2). The organic layer is then evaporated to dryness under nitrogen, and oxidized lipids in the residue are converted to PFB derivatives.

5.3. Extraction of oxidized lipids in the cells or tissues

Protein content is quantified by the Bradford assay using a portion of cell suspension or tissue homogenate for subsequent normalization of lipid levels. The cell suspension (typically 10^6 cells) or tissue homogenate (2 mg) is transferred to a cleaned glass tube containing chloroform and methanol (2:1, v/v, 5 ml). A mixture of 12 heavy isotope internal standards—[^{2}H$_4$]-9(S)-HODE, [^{2}H$_4$]-13(S)-HODE, [^{2}H$_8$]-5(S)-HETE, [^{2}H$_8$]-12(S)-HETE, [^{2}H$_8$]-15(S)-HETE, [^{2}H$_4$]-PGE$_2$, [^{2}H$_4$]-PGD$_2$, and [^{2}H$_4$]-LTB$_4$ (1 ng each), and [^{2}H$_6$]-20-HETE, [^{2}H$_4$]-PGF$_{2\alpha}$, [^{2}H$_4$]-11β-PGF$_2$, and [^{2}H$_4$]-8-iso-PGF$_{2\alpha}$ (10 ng each)—is then added to each sample. The samples are shaken for 15 min and centrifuged at 5000 rpm for 10 min. The supernatant from each tube is transferred to a new tube and washed with 1 ml of 0.9% NaCl solution. After vortex mixing and centrifugation to separate the two phases, the upper phase is removed. The steps for washing with 0.9% NaCl solution and separation are repeated and the combined lower phases are evaporated to dryness under nitrogen ready for PFB derivatization. Hydrolysis of esterified lipids is performed after dissolving the residue in methanol and chloroform (8:1, v/v, 850 μl) followed by the addition of 40% KOH (wt/v, 150 μl). After the tubes are filled with nitrogen, hydrolysis is conducted at 60° for 30 min. At the end of the hydrolysis, 50 mM phosphate buffer (pH 7.4, 700 μl) is added, and the pH of reaction mixture is adjusted with 150 μl of formic acid to pH 3 to 4. Nonesterified lipids are extracted with the solvent mixture of diethyl ether and hexane (1:1, v/v, 2 ml × 2), and the organic layer is then evaporated to dryness under nitrogen ready for PFB derivatization.

5.4. PFB derivatization

The final residue from extraction of the cell culture media, cells, or tissue homogenates is dissolved in 100 μl of acetonitrile followed by 100 μl PFB-Br in acetonitrile (1:19, v/v) and 100 μl of DIPE in acetonitrile (1:9, v/v). The solution is heated at 60° for 60 min, allowed to cool, evaporated to dryness under nitrogen at room temperature, and redissolved in 100 μl of hexane/ethanol (97:3, v/v). Analysis of the PFB derivatives by normal-phase chiral LC–electron-capture APCI/MRM/MS is conducted with a 20-μl aliquot of this solution using LC gradient-1.

5.5. Mass spectrometry of PFB derivatives

A Finnigan TSQ Quantum Ultra mass spectrometer (Thermo Fisher, San Jose, CA) equipped with an APCI source is used in the negative ion mode. For MRM analyses, unit resolution is maintained for both parent and product ions. Operating conditions are as follows: vaporizer temperature is 450°, heated capillary temperature is 250°, with a discharge current of 30 μA applied to the corona needle. Nitrogen is used for the sheath gas, auxiliary gas, and ion sweep gas set at 25, 3, and 3 (in arbitrary units), respectively. CID is performed using argon as the collision gas at 1.5 mTorr in the second (rf-only) quadrupole. An additional dc offset voltage is applied to the region of the second multipole ion guide (Q0) at 10 V to impart enough translational kinetic energy to the ions so that solvent adduct ions dissociate to form sample ions.

Targeted chiral LC-ECAPCI/MRM/MS analysis is conducted using PFB derivatives of 25 lipids and 12 heavy isotope analogue internal standards. The following MRM transitions are used: 9- and 13-oxo-ODE-PFB, m/z 293 → 113 (collision energy, 21 eV); 15-oxo-ETE-PFB, m/z 317 → 273 (collision energy, 14 eV); 13(R)- and 13(S)-HODE-PFB, m/z 295 → 195 (collision energy, 18 eV); [^{2}H$_4$]-13(S)-HODE-PFB, m/z 299 → 198 (collision energy, 18 eV); 9(R)- and 9(S)-HODE-PFB, m/z 295 → 171 (collision energy, 18 eV); [^{2}H$_4$]-9(S)-HODE-PFB, m/z 299 → 172 (collision energy, 18 eV); 20-HETE-PFB, m/z 319 → 289 (collision energy, 18 eV); [^{2}H$_6$]-20-HETE-PFB, m/z 325 → 295 (collision energy, 18 eV); 5(R)- and 5(S)-HETE-PFB, m/z 319 → 115 (collision energy, 15 eV); [^{2}H$_8$]-5(S)-HETE-PFB, m/z 327 → 116 (collision energy, 15 eV); 12(R)- and 12(S)-HETE-PFB, m/z 319 → 179 (collision energy, 14 eV); [^{2}H$_8$]-12(S)-HETE-PFB, m/z 327 → 184 (collision energy, 14 eV); 15(R)- and 15(S)-HETE-PFB, m/z 319 → 219 (collision energy, 13 eV); [^{2}H$_8$]-15(S)-HETE-PFB, m/z 327 → 226 (collision energy, 13 eV); 11(R)- and 11(S)-HETE-PFB, m/z 319 → 167 (collision energy, 16 eV); 8(R)- and 8(S)-HETE-PFB, m/z 319 → 155 (collision energy, 16 eV); LTB$_4$-PFB, m/z 335 → 195 (collision energy, 18 eV); [^{2}H$_4$]-LTB$_4$-PFB, m/z 339 → 197 (collision energy, 18 eV); PGE$_2$-PFB, PGD$_2$-PFB, 8-iso-PGE$_2$-PFB, m/z 351 → 271 (collision energy, 18 eV); [^{2}H$_4$]-PGE$_2$-PFB, [^{2}H$_4$]-PGD$_2$-PFB, m/z 355 → 275 (collision energy, 18 eV); 11β-PGF$_2$-PFB, PGF$_{2\alpha}$-PFB, 8-iso-PGF$_{2\alpha}$-PFB, m/z 353 → 309 (collision energy, 18 eV); and [^{2}H$_4$]-11β-PGF$_2$-PFB, [^{2}H$_4$]-PGF$_{2\alpha}$-PFB, [^{2}H$_4$]-8-iso-PGF$_{2\alpha}$-PFB m/z 357 → 313 (collision energy, 18 eV). Structures of selected PFB derivatives are shown in Fig. 9.4.

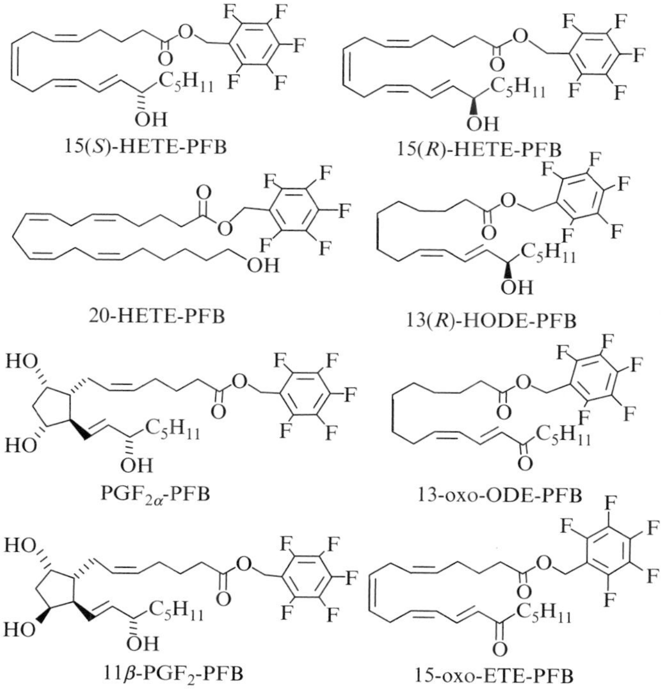

Figure 9.4 Structures of selected AA and LA-derived oxidized lipid PFB–ester derivatives.

5.6. Chiral liquid chromatography of PFB derivatives

The use of chiral chromatography with ethanol/hexane as solvent in combination with LC-electron capture APCI/MRM/MS for the PFB-ester derivatives of 9-HODE, 13-HODE, 11–HETE, 12-HETE, and 15-HETE resulted in separation of all the regioisomers. The PFB-esters of 9-HODE and 13-HODE are distinguished by the presence of product ions at m/z 171 and 195, respectively; whereas, 5-HETE, 12-HETE, 15-HETE, 11-HETE, and 8-HETE are distinguished by specific product ions at m/z 115, 179, 219, 167, and 155, respectively. The R-enantiomers generally elute ahead of the S-enantiomers. Unfortunately, the 12-HODE enantiomers do not separate to baseline and the 9-HODE enantiomers cannot be separated using this system (Lee *et al.*, 2003). With isopropanol/hexane as solvent the S-enantiomers tend to elute ahead of the R-enantiomers. The 9-HODE

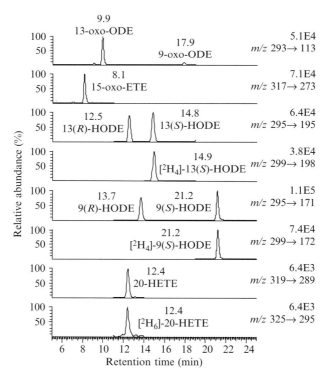

Figure 9.5 Chiral LC-MRM/ECAPCI/MS chromatograms from the analysis of 50 pg each of 13-oxo-ODE, 9-oxo-ODE, 15-oxo-ETE, 13(R)-HODE, 13(S)-HODE, 9(R)-HODE, and 9(S)-HODE as their PFB-ester derivatives, and 20-HETE together with [²H₄]-13(S)-HODE, [²H₄]-9(S)-HODE, and [²H₆]-20-HETE internal standards as their PFB-ester derivatives. A linear gradient elution with hexane, isopropanol, and methanol was performed (gradient-1).

enantiomers are now separated but the 15–HETE enantiomers are no longer resolved (Lee *et al.*, 2003). By using a linear gradient and a combination of methanol and isopropanol with hexane as solvent (gradient–1 below), all of the HODE and HETE isomers and enantiomers can be separated (Figs. 9.5 and 9.6). Isomeric PG derivatives such as PGD_2 and PGE_2, as well as PG and isoP isomers such as $PGF_{2\alpha}$ and 8–*iso*-$PGF_{2\alpha}$, are also resolved under these conditions (Fig. 9.7). Normal-phase chiral ECAPCI/MS analysis is conducted using a Waters Alliance 2690 HPLC system (Waters Corp., Milford, MA). A Chiralpak AD-H column (250 × 4.6 mm i.d., 5 μm; Daicel Chemical Industries, Ltd., Tokyo, Japan) is employed for gradient–1 with a flow rate of 1.0 ml/min. Solvent A is hexane and solvent B is methanol/isopropanol (1:1, v/v). Gradient–1 is run in the linear mode as follows: 2% B at 0 min, 2% B at 3 min, 3.6% B at 11 min, 8% B at 15 min, 8% B at 27 min, 50% B at 30 min, 50% B at 35 min, and 2% B at 37 min. Post-column addition of methanol (0.75 ml/min) is employed to avoid the formation of

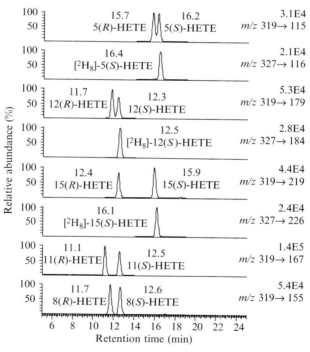

Figure 9.6 Chiral LC-MRM/ECAPCI/MS chromatograms from the analysis of 50 pg each of 5(*R*)-HETE, 5(*S*)-HETE, 12(*R*)-HETE, 12(*S*)-HETE, 15(*R*)-HETE, 15(*S*)-HETE, 11(*R*)-HETE, 11(*S*)-HETE, 8(*R*)-HETE, and 8(*S*)-HETE as their PFB-ester derivatives, together with [$^2H_8$]-5(*S*)-HETE, [$^2H_8$]-12(*S*)-HETE, and [$^2H_8$]-15(*S*)-HETE internal standards as their PFB-ester derivatives. A linear gradient elution with hexane, isopropanol, and methanol was performed (gradient-1).

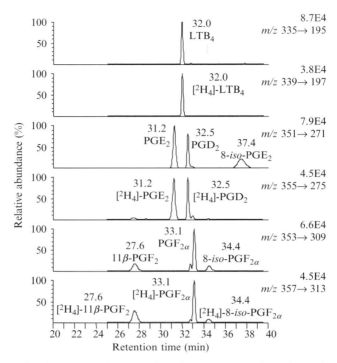

Figure 9.7   Chiral LC-MRM/ECAPCI/MS chromatograms from the analysis of 50 pg each of LTB$_4$, PGE$_2$, PGD$_2$, 8-*iso*-PGE$_2$, 11β-PGF$_2$, PGF$_{2\alpha}$, and 8-*iso*-PGF$_{2\alpha}$ as their PFB-ester derivatives, together with [^{2}H$_4$]-LTB$_4$, [^{2}H$_4$]-PGE$_2$, [^{2}H$_4$]-PGD$_2$, [^{2}H$_4$]-11β-PGF$_2$, [^{2}H$_4$]-PGF$_{2\alpha}$, and [^{2}H$_4$]-8-*iso*-PGF$_{2\alpha}$ internal standards as their PFB-ester derivatives. A linear gradient elution with hexane, isopropanol, and methanol was performed (gradient-1).

black deposit on top of the corona needle, which causes the decrease of sensitivity. Separations are performed at 30° (see Figs. 9.5 through 9.7).

5.7. Standard curves and quantitative analysis of oxidized lipids

Calibration standard samples are prepared in the same way as regular samples, substituting media with serum-free media (3 ml), and cells or tissues with the solvent mixture of chloroform and methanol (2:1, v/v, 5 ml), respectively. These matrices are spiked with the following amounts of 25 authentic lipid standards: 10, 20, 50, 100, 200, 500, 1000 pg, and a mixture of internal standards—[^{2}H$_4$]-9(S)-HODE, [^{2}H$_4$]-13(S)-HODE, [^{2}H$_8$]-5(S)-HETE, [^{2}H$_8$]-12(S)-HETE, [^{2}H$_8$]-15(S)-HETE, [^{2}H$_4$]-PGE$_2$, [^{2}H$_4$]-PGD$_2$, [^{2}H$_4$]-LTB$_4$ (1 ng each), and [^{2}H$_6$]-20-HETE, [^{2}H$_4$]-PGF$_{2\alpha}$, [^{2}H$_4$]-11β-PGF$_2$, and [^{2}H$_4$]-8-*iso*-PGF$_{2\alpha}$ (10 ng each). Oxidized lipids are then extracted, purified, derivatized, and analyzed as described above for the analytical samples. Calibration curves are obtained with a linear regression analysis of peak area ratios of analytes against internal standard. Typical regression lines for 11(*R*)-HETE, 15(*R*)-HETE, 15(*S*)-HETE, and PGE$_2$ in the cell culture medium are y = 0.0029x + 0.0169 (r^2 = 0.9924), y = 0.0099x + 0.0459 (r^2 = 0.9989), y = 0.0015x + 0.0421 (r^2 = 0.9570), y = 0.0030x + 0.016 (r^2 = 0.9915), and y = 0.0026x + 0.0044 (r^2 = 0.9988), respectively. Concentrations of bioactive lipids are calculated by interpolation from the calculated regression lines.

6. SUMMARY

Recent developments in APCI instrumentation as exemplified by the Finnigan TSQ Ultra AM mass spectrometer have improved the sensitivity of ECAPCI/MS compared to that obtained in our original studies (Singh *et al.*, 2000). In particular, the close proximity to the corona discharge to the sampling cone in the source of the mass spectrometer results in more efficient electron capture and increased sensitivity. Furthermore, a much higher corona discharge current (30 μA) can be used when compared earlier instrumentation, which has increased the efficiency of electron capture and improved sensitivity. The use of higher corona discharge currents tends to cause a black deposit to build up in the source. However, this can be prevented by the post-column addition of 0.75 ml/min of methanol. Using a corona discharge current of 30 μA, it is now possible to quantify as little as 100 fg of HETEs and HODEs during a targeted lipidomics analysis.

The high sensitivity and specificity of ECAPCI/MS has proved to be particularly useful in our studies on the COX-2-mediated metabolism of AA. We were able to show the importance of endogenous 15(*S*)-HPETE formation in unstimulated rat intestinal epithelial cells as reflected by 15(*S*)-HETE concentrations (Lee *et al.*, 2005b). Previous studies had shown that the heptanone-etheno-2′-deoxyguanosine (HEdGuo) DNA–adduct (Rindgen *et al.*, 1999) arose from the homolytic decomposition of 15(*S*)-HPETE to ONE (Lee and Blair, 2000; Williams *et al.*, 2005). Therefore, HEdGuo found in the DNA of rat intestinal epithelial cells was thought to arise from COX-2–mediated AA metabolism to 15(*S*)-HPETE. This was confirmed by the observation that a selective COX-2 inhibitor prevented both 15(S)-HETE and HEdGuo formation (Lee *et al.*, 2005b). The vitamin C–mediated increase in HEdGuo DNA adducts and the decrease in 15(S)-HETE concentrations was in keeping with the concept that vitamin C can induce the homolytic decomposition of 15(*S*)-HPETE to ONE (Lee *et al.*, 2001; Williams *et al.*, 2005). Increased HEdGuo formation was also observed in intestinal polyps from min mice, a mouse model in which COX-2 is upregulated (Williams *et al.*, 2006).

More recently, we have used the ECAPCI technique to show that 15-oxo-ETE is a COX-2–mediated metabolite of AA, and that it arises through the oxidation of 15(S)-HETE (Lee *et al.*, 2007). Other groups have used ECAPCI/MS for the separation and quantification of complex mixtures of isoPs (Yin *et al.*, 2005b). It is noteworthy that LC-ECAPCI/MS methodology can be readily modified to include oxidized lipids derived from PUFAs other than AA and LA such as eicosapentaenoic acid and docosahexaenoic acid. For example, it was used to establish the regiochemistry of neuroprostanes generated from the peroxidation of docosahexaenoic acid *in vitro* and *in vivo* (Yin *et al.*, 2005a). Finally, we anticipate that LC-ECAPCI/MS will make a significant contribution to future global lipidomics studies (Watson, 2006) as well as in the development of quantitative analysis of lipid-derived biomarkers of toxicity and disease (Ackermann *et al.*, 2006; Kamel and Prakash, 2006).

ACKNOWLEDGMENTS

Supported by National Institutes of Health grants RO1 CA95586 and P30 ES013508.

REFERENCES

Ackermann, B. L., Hale, J. E., and Duffin, K. L. (2006). The role of mass spectrometry in biomarker discovery and measurement. *Curr. Drug Metab.* **7,** 525–539.

Barrow, S. E., Ward, P. S., Sleightholm, M. A., Ritter, J. M., and Dollery, C. T. (1989). Cigarette smoking: Profiles of thromboxane- and prostacyclin-derived products in human urine. *Biochim. Biophys. Acta* **993,** 121–127.

Blair, I. A. (1990). Electron-capture negative-ion chemical ionization mass-spectrometry of lipid mediators. *Methods Enzymol.* **187,** 13–23.

Blair, I. A. (2001). Lipid hydroperoxide-mediated DNA damage. *Exp. Gerontol.* **36,** 1473–1481.

Blair, I. A. (2005). Mass spectrometry approaches to elucidate the role of oxidative stress in cancer and toxicology. *In* "Encyclopedia of Mass Spectrometry" (R. M. Caprioli and M. L. Gross, eds.). Elsevier, New York.

Blair, I. A. (2006). Endogenous glutathione adducts. *Curr. Drug Metab.* **7,** 853–872.

Blair, I. A., Barrow, S. E., Waddell, K. A., Lewis, P. J., and Dollery, C. T. (1982). Prostacyclin is not a circulating hormone in man. *Prostaglandins* **23,** 579–589.

Brash, A. R., Boeglin, W. E., and Chang, M. S. (1997). Discovery of a second 15S-lipoxygenase in humans. *Proc. Natl. Acad. Sci. USA* **94,** 6148–6152.

Burinsky, D. J., Williams, J. D., Thornquest, A. D., Jr., and Sides, S. L. (2001). Mass spectral fragmentation reactions of a therapeutic 4-azasteroid and related compounds. *J Am. Soc. Mass Spectrom.* **12,** 385–398.

Capdevila, J. H., and Falck, J. R. (2002). Biochemical and molecular properties of the cytochrome P450 arachidonic acid monooxygenases. *Prostaglandins Other Lipid Mediat.* **68–69,** 325–344.

Capdevila, J. H., Harris, R. C., and Falck, J. R. (2002). Microsomal cytochrome P450 and eicosanoid metabolism. *Cell Mol. Life Sci.* **59**, 780–789.

Carroll, D. I., Dzidic, I., Stillwell, R. N., Haegele, K. D., and Horning, E. C. (1975). Atmospheric-pressure ionization mass-spectrometry – corona discharge ion-source for use in liquid chromatograph mass spectrometer-computer analytical system. *Anal. Chem.* **47**, 2369–2373.

Chen, X., Sood, S., Yang, C. S., Li, N., and Sun, Z. (2006). Five-lipoxygenase pathway of arachidonic acid metabolism in carcinogenesis and cancer chemoprevention. *Curr. Cancer Drug Targets* **6**, 613–622.

Giese, R. W. (2000). Electron-capture mass spectrometry: Recent advances. *J. Chromatogr. A* **892**, 329–346.

Hayen, H., Jachmann, N., Vogel, M., and Karst, U. (2002). LC-electron capture APCI-MS for the determination of nitroaromatic compounds. *Analyst* **127**, 1027–1030.

Higashi, T., Takido, N., Yamauchi, A., and Shimada, K. (2002). Electron-capturing derivatization of neutral steroids for increasing sensitivity in liquid chromatography/negative atmospheric pressure chemical ionization-mass spectrometry. *Anal. Sci.* **18**, 1301–1307.

Horning, E. C., Carroll, D. I., Dzidic, I., Haegele, K. D., Horning, M. G., and Stillwel, R. N. (1974). Atmospheric-pressure ionization (API) mass-spectrometry—solvent-mediated ionization of samples introduced in solution and in a liquid chromatograph effluent stream. *J. Chromatogr. Sci.* **12**, 725–729.

Hunt, D. F., Stafford, G. C., Crow, F. W., and Russell, J. W. (1976). Pulsed positive negative-Ion chemical ionization mass-spectrometry. *Anal. Chem.* **48**, 2098–2105.

Jones, R., del-Alvarez, L. A., Alvarez, O. R., Broaddus, R., and Das, S. (2003). Arachidonic acid and colorectal carcinogenesis. *Mol. Cell. Biochem.* **253**, 141–149.

Kamel, A., and Prakash, C. (2006). High performance liquid chromatography/atmospheric pressure ionization/tandem mass spectrometry (HPLC/API/MS/MS) in drug metabolism and toxicology. *Curr. Drug Metab.* **7**, 837–852.

Kamitani, H., Geller, M., and Eling, T. (1999). The possible involvement of 15-lipoxygenase/leukocyte type 12-lipoxygenase in colorectal carcinogenesis. *Adv. Exp. Med. Biol.* **469**, 593–598.

Lee, S. H.., and Blair, I. A. (2000). Characterization of 4-oxo-2-nonenal as a novel product of lipid peroxidation. *Chem. Res. Toxicol.* **13**, 698–702.

Lee, S. H., and Blair, I. A. (2001). Oxidative DNA damage and cardiovascular disease. *Trends Cardiovasc. Med.* **11**, 148–155.

Lee, S. H., Oe, T., and Blair, I. A. (2001). Vitamin C–induced decomposition of lipid hydroperoxides to endogenous genotoxins. *Science* **292**, 2083–2086.

Lee, S. H., Williams, M. V., and Blair, I. A. (2005a). Targeted chiral lipidomics analysis. *Prostaglandins Other Lipid Mediat.* **77**, 141–157.

Lee, S. H., Williams, M. V., Dubois, R. N., and Blair, I. A. (2003). Targeted lipidomics using electron capture atmospheric pressure chemical ionization mass spectrometry. *Rapid Commun. Mass Spectrom.* **17**, 2168–2176.

Lee, S. H., Williams, M. V., Dubois, R. N., and Blair, I. A. (2005b). Cyclooxygenase-2–mediated DNA damage. *J. Biol. Chem* **280**, 28337–28346.

Lee, S. H., Rangiah, K., Williams, M. V., Wehr, A. Y., Dubois, R. N., and Blair, I. A. (2007). Cyclooxygenase-2–mediated metabolism of arachidonic acid to 15-oxo-eicosatetraenoic acid by rat intestinal epithelial cells. *Chem. Res. Toxicol.* In press.

Matuszewski, B. K. (2006). Standard line slopes as a measure of a relative matrix effect in quantitative HPLC-MS bioanalysis. *J Chromatogr. B Analyt. Technol. Biomed. Life Sci.* **830**, 293–300.

Milne, G. L., Musiek, E. S., and Morrow, J. D. (2005). F2-isoprostanes as markers of oxidative stress *in vivo* an overview. *Biomarkers* **10**(Suppl. 1), S10–S23.

Montine, T. J., and Morrow, J. D. (2005). Fatty acid oxidation in the pathogenesis of Alzheimer's disease. *Am. J Pathol.* **166**, 1283–1289.

Morrow, J. D., Zackert, W. E., Yang, J. P., Kurhts, E. H., Callewaert, D., Dworski, R., Kanai, K., Taber, D., Moore, K., Oates, J. A., and Roberts, L. J. (1999). Quantification of the major urinary metabolite of 15-F2t-isoprostane (8-iso-PGF2alpha) by a stable isotope dilution mass spectrometric assay. *Anal. Biochem.* **269**, 326–331.

Musiek, E. S., Yin, H., Milne, G. L., and Morrow, J. D. (2005). Recent advances in the biochemistry and clinical relevance of the isoprostane pathway. *Lipids* **40**, 987–994.

Porter, N. A., Caldwell, S. E., and Mills, K. A. (1995). Mechanisms of free radical oxidation of unsaturated lipids. *Lipids* **30**, 277–290.

Ramesha, C. S., and Pickett, W. C. (1986). Measurement of sub-picogram quantities of platelet activating factor (AGEPC) by gas chromatography/negative ion chemical ionization mass spectrometry. *Biomed. Environ. Mass Spectrom.* **13**, 107–111.

Rindgen, D., Nakajima, M., Wehrli, S., Xu, K., and Blair, I. A. (1999). Covalent modifications to 2'-deoxyguanosine by 4-oxo-2-nonenal, a novel product of lipid peroxidation. *Chem. Res. Toxicol.* **12**, 1195–1204.

Rokach, J., Kim, S., Bellone, S., Lawson, J. A., Praticó, D., Powell, W. S., and Fitzgerald, G. A. (2004). Total synthesis of isoprostanes: Discovery and quantitation in biological systems. *Chem. Phys. Lipids* **128**, 35–56.

Sharma, R. A., Manson, M. M., Gescher, A., and Steward, W. P. (2001). Colorectal cancer chemoprevention: Biochemical targets and clinical development of promising agents. *Eur. J. Cancer* **37**, 12–22.

Singh, G., Gutierrez, A., Xu, K., and Blair, I. A. (2000). Liquid chromatography/electron capture atmospheric pressure chemical ionization/mass spectrometry: Analysis of penta-fluorobenzyl derivatives of biomolecules and drugs in the attomole range. *Anal. Chem.* **72**, 3007–3013.

Strife, R. J., and Murphy, R. C. (1984). Stable isotope labelled 5-lipoxygenase metabolites of arachidonic acid: Analysis by negative ion chemical ionization mass spectrometry. *Prostaglandins Leukot. Med.* **13**, 1–8.

Teismann, P., Vila, M., Choi, D. K., Tieu, K., Wu, D. C., Jackson-Lewis, V., and Przedborski, S. (2003). COX-2 and neurodegeneration in Parkinson's disease. *Ann. N. Y. Acad. Sci.* **991**, 272–277.

Wang, D., and Dubois, R. N. (2006). Prostaglandins and cancer. *Gut* **55**, 115–122.

Wang, D., Wang, M., Cheng, Y., and Fitzgerald, G. A. (2005). Cardiovascular hazard and non-steroidal anti-inflammatory drugs. *Curr. Opin. Pharmacol.* **5**, 204–210.

Watson, A. D. (2006). Lipidomics: A global approach to lipid analysis in biological systems. *J. Lipid Res.* **47**, 2101–2111.

Williams, M. V., Lee, S. H., and Blair, I. A. (2005). Liquid chromatography/mass spectrometry analysis of bifunctional electrophiles and DNA adducts from vitamin C mediated decomposition of 15-hydroperoxyeicosatetraenoic acid. *Rapid Commun. Mass Spectrom.* **19**, 849–858.

Williams, M. V., Lee, S. H., Pollack, M., and Blair, I. A. (2006). Endogenous lipid hydroperoxide-mediated DNA-adduct formation in min mice. *J. Biol. Chem.* **281**, 10127–10133.

Yin, H., and Porter, N. A. (2005). New insights regarding the autoxidation of polyunsaturated fatty acids. *Antioxid. Redox. Signal.* **7**, 170–184.

Yin, H. Y., Musiek, E. S., Gao, L., Porter, N. A., and Morrow, J. D. (2005a). Regiochemistry of neuroprostanes generated from the peroxidation of docosahexaenoic acid *in vitro* and *in vivo*. *J. Biol. Chem.* **280**, 26600–26611.

Yin, H. Y., Porter, N. A., and Morrow, J. D. (2005b). Separation and identification of F-2-isoprostane regioisomers and diastereomers by novel liquid chromatographic/mass spectrometric methods. *J. Chromatogr. B Analyt. Technol. Biomed. Life Sci.* **827**, 157–164.

CHAPTER TEN

SHOTGUN LIPIDOMICS BY TANDEM MASS SPECTROMETRY UNDER DATA-DEPENDENT ACQUISITION CONTROL

Dominik Schwudke,* Gerhard Liebisch,[†] Ronny Herzog,* Gerd Schmitz,[†] *and* Andrej Shevchenko*

Contents

Abstract

Data-dependent acquisition of full MS/MS spectra from all detectable (or, alternatively, preselected) lipid precursors produces a rich data set, whose subsequent interpretation by the dedicated software LipidInspector emulates the simultaneous acquisition of an unlimited number of precursor and neutral loss scans in a single analysis. Using logical operations, emulated scans can be combined into highly specific data interpretation routines (termed Boolean

* MPI of Molecular Cell Biology and Genetics, Dresden, Germany
[†] Institute of Clinical Chemistry and Laboratory Medicine, University of Regensburg, Regensburg, Germany

Methods in Enzymology, Volume 433
ISSN 0076-6879, DOI: 10.1016/S0076-6879(07)33010-3

scans) enabling in-depth structural characterization of fragmented precursors. Alternatively, a small number of preselected precursors can be fragmented regardless of their relative intensities in survey spectra, hence emulating selected reaction monitoring (SRM) analysis that attains both high detection specificity and sensitivity. Although the data-dependent acquisition approach is, in principle, cross-platform, it benefits from the high mass resolution capacity of hybrid tandem mass spectrometers with time-of-flight and, especially, Fourier transform or Orbitrap analyzers.

1. INTRODUCTION

In shotgun lipidomics (Ekroos *et al.*, 2003; Han and Gross, 2005), total lipid extracts are infused directly into a mass spectrometer via the electrospray ion source and lipid species are subsequently identified and quantified by tandem mass spectrometry using lipid class–specific and (or) lipid species–specific precursor ion scans (PIS) and neutral loss scans (NLS), reviewed in Han and Gross, 2003, 2005; Pulfer and Murphy, 2003; and Wenk, 2005. By varying the solvent composition and spraying conditions, it is possible to specifically enhance the ionization of certain lipid classes and, hence, to improve the dynamic range of lipid detection (Han *et al.*, 2006). PIS and NLS are typical features of triple quadrupole mass spectrometers, the workhorses of the lipidomics field. However, these instruments can only acquire one PIS or NLS spectrum at a time. When quantitative profiling of multiple lipid classes requires several PIS and NLS, the analysis is successively repeated several times (Brugger *et al.*, 1997). In contrast, quadrupole time–of–flight (QqTOF) mass spectrometers can acquire multiple precursor ion spectra in parallel, which, together with the accurate selection of m/z of fragment ions, opens up interesting analytical opportunities (Ejsing *et al.*, 2006a; Ekroos *et al.*, 2002). It is known that collision–induced dissociation (CID) of molecular anions of glycerophospholipid species produces abundant acyl anion fragments of their fatty acid moieties. Therefore, the interpretation of precursor ion spectra, simultaneously acquired for a multitude of plausible acyl anion fragments, could identify and quantify individual molecular species (Ejsing *et al.*, 2006a).

However, QqTOF machines can only acquire multiple precursor ion spectra, but not neutral loss spectra. The absence of NLS hampers profiling of several glycerophospholipid classes, especially in positive ion mode. Once acquired, the data set of precursor ion spectra is "frozen," and is not amenable for further reprocessing or manual interpretation by concidering other fragment ions produced in alternative CID pathways. We note, however, that the mass resolution and accuracy of emerging hybrid instruments with Fourier transform (Syka *et al.*, 2004) or Orbitrap (Makarov *et al.*, 2006) mass analyzers enable unequivocal determination of the elemental composition of fragment ions solely by their accurately measured masses, hence facilitating the in-depth structural characterization of lipid molecules (Ejsing *et al.*, 2006b). Therefore, the acquisition and interpretation of complete high-resolution MS/MS spectra (rather than their subsets, effectively used in PIS) offers important analytical advantages. Because of the elemental composition constraints, masses of plausible lipid precursors occupy well-spaced m/z slots, rather than populating continuously the entire m/z range. Hence, in many instances, spectra acquisition in scanning mode would be, in fact, impractical.

Here we demonstrate that data-dependent acquisition of full MS/MS spectra from all detectable (or, alternatively, preselected) precursors provides a rich lipidomics data set amenable to the versatile interpretation in a user-defined manner (Schwudke *et al.*, 2006). Rapid extraction of intensities of relevant fragment ions out of each spectrum from the complete MS/MS data set effectively emulates the unlimited number of precursor and neutral loss scans, without compromising the accuracy of identification and quantification of lipid species. Furthermore, in the course of postacquisition MS/MS spectra processing, logical operations could combine emulated scans into lipid structure–specific data interpretation routines, known as Boolean scans.

Alternatively, the inclusion list of plausible precursors could be collapsed down to a very few masses of interest. Hence, MS/MS experiments followed by the extraction of intensities of specific fragment ions from corresponding *dta* files, would effectively emulate the method of selected reaction monitoring (SRM) that dramatically enhances the detection sensitivity and specificity (Liebisch *et al.*, 2006). Rapid and flexible adjustment of data acquisition and processing routines that are tailored to the specific goals of a particular experiment would expand the gamut of generic analytical tools in lipidomics.

2. Procedure

2.1. Sample preparation for mass spectrometric analysis

Stock solutions of lipid standards were prepared in the specified concentrations in $CHCl_3/MeOH/2$-propanol 1:2:4 (v/v/v) containing 7.5-mM ammonium acetate. Before the analysis, samples were vortexed thoroughly and centrifuged for 5 minutes at 14,000×g (14,000 rpm) on a Minispin centrifuge (Eppendorf, Hamburg, Germany). Samples were loaded onto 96-well plates and sealed with aluminum foil.

2.2. Lipid extraction

Lipids were extracted according to Bligh and Dyer (Bligh and Dyer, 1959). Briefly, three volumes of methanol:chloroform = 2:1 (v/v) were added to 0.8 (v/v) volume aqueous samples. To achieve phase separation, 1 volume of water and 1 volume of chloroform were added, and the chloroform phase separated upon centrifugation was collected and dried. For quantitative analysis, non-naturally occurring lipid species were added as internal standards before extraction.

Where specified, total extracts were separated by TLC, and scrapped bands were extracted and analyzed as described (Schwudke *et al.*, 2006).

2.3. Cholesterol recovery and derivatization

Free cholesterol was converted into cholesteryl acetate as described previously (Liebisch *et al.*, 2006). Briefly, a 1:5 (v/v) mixture of acetyl chloride: chloroform was added to dried lipid extracts, and the solutions were incubated for 60 min at room temperature. Under these conditions, no transesterification of naturally occurring cholesteryl ester species occurred. For the quantification of cholesterol, internal standards of [25,26,26,26,27,27,27-D_7]-cholesterol and two non-naturally occurring cholesteryl ester species—heptadecanoate (CE 17:0) and behenate (CE 22:0)—were added to analyzed samples before lipid extraction.

2.4. Mass spectrometers

Lipid extracts were analyzed on a modified QSTAR Pulsar *i* quadrupole time-of-flight mass spectrometer (MDS Sciex, Concord, Canada) equipped with a robotic nanoflow ion source NanoMate HD (Advion BioSciences, Ithaca, NY). The instrument was calibrated in MS/MS mode using a synthetic lipid standard 1-palmitoyl-2-docosahexaenoyl-*sn*-glycero-3-phosphocholine as previously described (Ekroos and Shevchenko, 2002). Analytical quadrupole Q1 was calibrated according to the instructions from the manufacturer. Its resolution was adjusted such that, from selected precursors, either entire isotopic cluster (low-resolution settings) or only the monoisotopic ions (unit resolution settings), were transmitted. Mass resolution offsets were saved as separate settings tables. MS/MS experiments were performed at the collision energy offset of 40 eV. Ionization voltage at the NanoMate source was set to 950 V, gas pressure to 1.25 psi, and the source was controlled by Chipsoft (v. 6.3.2; Advion BioSciences). Lipid extracts were infused at the flow rate of *ca* 250 nl/min. A typical sample volume of 10 μl allowed more than 40 min of stable electro-spray time. The spraying stability was monitored by the total ion count (TIC) signal reported from survey TOF MS spectra in each DDA cycle.

Acetylated cholesterol and cholesteryl esters were quantified on a Quattro Ultima triple-quadrupole mass spectrometer (Micromass, Manchester, UK) by direct-flow injection analysis using a HTS PAL autosampler (Zwingen, Switzerland) and an Agilent 1100 binary pump (Waldbronn, Germany) with a solvent mixture of methanol containing 10 mM ammonium acetate and chloroform (3:1, v/v). A flow gradient was performed starting at 55 μl/min for 6 s followed by 30 μl/min for 1.0 min and then increased to 250 μl/min for another 12 s. The triple quadrupole mass spectrometer was equipped with an electro-spray ion source operated in positive ion mode under the following settings: capillary voltage 3.5 kV, cone voltage 50 V, collision energy 13 eV, collision gas argon, and collision gas pressure 1 mTorr. Both Q1 and Q3 quadrupoles were operated under better than unit mass resolution.

2.5. Data-dependent acquisition setup

In our experiments, data-dependent acquisition (DDA; also called information-dependent acquisition [IDA]) on a QSTAR mass spectrometer was controlled by Analyst QS (v. 1.1; Applied Biosystems). Depending on the targeted lipid classes and, respectively, required methods of post-acquisition processing of MS/MS spectra, DDA cycles could be set up in many different ways. A generic DDA cycle that provides MS/MS data for emulating both precursor and neutral loss scans typically includes one TOF MS survey scan for the time period of 2 s followed by two successive MS/MS experiments (each of 10 to 30 s) that target the same precursor ion. In the first MS/MS experiment, the analytical quadrupole Q1 is operated under the unit resolution settings and only transmits monoisotopic peaks of the fragmented precursors. At the same time, the TOF analyzer is set to detect low-molecular-weight fragments. Note that narrowing TOF m/z range improves the duty cycle (Chernushevich et al., 2001) and, consequently, sensitivity. Subsequently, this spectrum will be taken for emulating precursor ion scans. In the next MS/MS experiment, Q1 is operated under the low resolution settings and the entire isotopic cluster of the fragmented precursor is transmitted. The TOF analyzer acquires spectra in higher m/z range, usually up to the precursor m/z (assuming that precursor ions are singly charged). Fragments detected in the second cycle are mostly used for emulating neutral loss scans. Note that in neutral loss scans, partial co-isolation of neighboring precursors of the same lipid class (e.g., of the two species that differ by one double bond in a fatty acid moiety and, hence, are spaced by 2 Da) does not affect quantification accuracy (Schwudke et al., 2006). However, low mass resolution of Q1 improves ion transmittance, and, hence, enhances the analysis sensitivity. Upon completing the acquisition cycle, the instrument starts a new cycle, in which either the next precursor candidate is selected by considering the intensities of peaks in the survey scan, or another precursor m/z is taken from the inclusion list. The m/zs of already fragmented precursors are excluded until the end of the analysis. It is always worth to include TOF MS survey scan in the DDA cycle, even if it is not required for the subsequent acquisition of MS/MS spectra. Recording TIC from the survey scan at each cycle helps to monitor ESI spray stability throughout the entire DDA experiment, which is essential for robust quantitative determinations. Once the experiment is completed, the acquired MS/MS spectra are exported as individual files in *dta* format. All *dta*s produced in the same analysis are collected into one data set folder for subsequent interpretation.

Because of much slower data acquisition speed, DDA on triple quadrupole mass spectrometers is usually navigated by inclusion lists, in which precursor m/z are either precalculated or deduced from rapid, low-resolution, lipid class–specific PIS or NLS. If samples of limited volumes are infused into a mass spectrometer at 10- to 50-μl/min flow rate, TIC does not reach a plateau, but rather is shaped as a peak. All scans above the half-height of the TIC are then averaged, and the centroided spectrum is produced from the combined continuum spectra. The centroided peak list is exported as a spreadsheet for correcting the peak intensities within partially overlapping isotopic clusters. The peaks detected at the signal-to-noise ratio above the value of 3, and matching the expected masses of species of the targeted lipid class are then compiled into inclusion list for selected reaction monitoring (SRM). The analysis in SRM mode is particularly useful in the quantification of low abundant lipid species. SRM spectra are processed as described above without conversion of continuum to centroided data. Quantification is performed using calibration plots produced by spiking analytes with non–natural synthetic lipid species used as internal standards (Liebisch *et al.*, 2004).

2.5.1. Inclusion lists for DDA-driven lipid profiling

Inclusion lists are important elements of DDA-driven lipidomic routines. Despite structural divergence of glycerophospholipids, species of different classes often have overlapping nominal masses, although their exact masses might differ. Because in MS/MS experiments on a QqTOF mass spectrometer Q1 quadrupole mass analyzer selects precursors with the unit or lower mass resolution, a single inclusion list serving all plausible glycerophospholipid precursors could be compiled. Inclusion lists target the analysis on interesting precursors, irrespectively of other peaks observed in survey scan spectra. This, effectively, balances the exploratory and focused strategies in lipidomics profiling.

Quantification of cholesterol and its esters serves as a good example of the focused lipidomics scenario. In this case, the selected m/zs are analyzed in SRM mode regardless of their intensity ratios to other peaks in the survey spectrum. Alternatively, the exploratory analysis could target all common glycerophospholipid precursors, or triacylglycerol (TAG) precursors, or ceramide precursors, among others. Once a survey scan spectrum is acquired, the listed precursors are ranked by the abundance of their detectable peaks and analyzed accordingly. Generic inclusion lists used by Schwudke *et al.* (2006) for DDA on a QSTAR Pulsar *i* quadrupole time-of-flight mass spectrometer comprised 100 precursor entries within the range of *m/z* 690 to 890 for glycerophospholipids and *m/z* 790 to 990 for TAGs.

3. AUTOMATED INTERPRETATION OF MS/MS SPECTRA ACQUIRED IN DDA EXPERIMENTS

A typical DDA experiment produces the pool of *dta* files, each of which represents a single MS/MS spectrum acquired from the selected lipid precursor. The first line of each *dta* file contains the precursor mass and charge, followed by a peak list of centroided *m/z* and intensities of detected fragments. All *dta* files acquired in the same experiment are processed by LipidInspector (available free of charge from Scionics Computer Innovations, at www.scionics.de). Depending on acquisition mode polarity, collision-induced dissociation of molecular ions of glycerophospholipids yields fragments of head groups and/or fatty acid moieties, either as ions or neutrals. LipidInspector uses the predefined list of *m/z* of these specific fragments and neutral losses. In each MS/MS spectrum, LipidInspector identifies fragments matching the masses of list entries with defined tolerance and, if requested, checks if they also meet other predefined selection criteria (Fig. 10.1).

Subsequently, LipidInspector evokes a naming routine that annotates the corresponding precursors considering their intact masses and matched fragments. We note that *m/z* of precursors are either determined in full TOF MS survey scans, or within 0.2-Da window around *m/z* taken from the inclusion list, Importantly, no residual precursor peaks detectable in the acquired MS/MS spectra are considered. Typically, glycerophospholipid precursors are designated with their lipid class (PE, PC, PG, and so on) and sum composition (the total number of carbon atoms and double bonds in both fatty acid moieties). LipidInspector reports the matched peaks along with their intensities as a tab-delimited text file, which could be directly opened in Microsoft Excel. For the quantitative analysis, isotopic correction of peak intensities is performed as described by Han and Gross (2005).

Note that the spectra obtained with low (unit or less) and high (over ~8000 FWHM) mass resolution are interpreted differently by LipidInspector. For example, in high-resolution spectra, it is possible to distinguish, within the same lipid class, species with two ester bonds and species having one

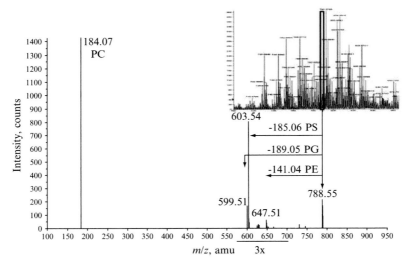

Figure 10.1 Mass spectrum of a total lipid extract acquired at the QSTAR mass spectrometer in TOF MS mode. MS/MS spectrum acquired from the precursor ion with m/z788.55 (framed in the upper pane) contains specific signature ions of several isobaric lipid species co-selected by the quadrupole analyzer. For example, the fragment with m/z184.07 was produced from the phosphocholine group of sphingomyelins and phosphatidylcholines, and the fragment with m/z 647.51 was produced by a neutral loss of phosphoethanolamine group from the head group of the PE precursor (other neutral losses are also annotated in the figure). For interpreting each spectrum, the LipidInspector program uses the list of characteristic masses and mass differences relative to $m=z$ of the precursor. Upon opening the spectrum, LipidInspector identifies those ions and, knowing the exact mass of the fragmented precursor, annotates lipid species by their lipid class and sum composition. Here the software concluded that the peak with m/z 788.55 isolated by Q1 and fragmented, in fact, stands for PC 36:1 (exact m/z 788.6163), PS 36:2 (m/z788.5436), PG 36:4 (m/z 788.5436), and PE 39:2 (m/z 788.6163). Intensities of corresponding fragment peaks were reported for the subsequent quantification of species. Note that because no chromatographic separation is involved, intensities of fragment peaks are characteristic of the concentrations of corresponding components and do not require integration in the time dimension.

ester and one ether bond, even if they are isobaric. The difference of 0.0364 Da ($m_O - m_{CH4}$) between their exact masses could be confidently determined in MS/MS mode because, compared to survey scans, chemical noise is much reduced, and it is also unlikely that specific fragments (contrary to precursors) overlap with other isobaric ions. Interestingly, hybrid instruments with FT or Orbitrap analyzers could reliably distinguish the corresponding species even in survey scans, if acquired at the target mass resolution exceeding 100,000 (FWHM), with no recourse to MS/MS (Schwudke, 2007).

In experiments on a triple quadrupole mass spectrometer, SRM spectra can be exported as peak lists and analyzed by LipidInspector or by dedicated macros in Microsoft Excel (Liebisch *et al.*, 2004).

4. BOOLEAN SCANS: A NOVEL SCAN TYPE ENABLED BY POSTACQUISITION DATA PROCESSING

Precursor ion scans and neutral loss scans effectively identify molecules that, upon CID, produce either the same fragment, or fragments with the same mass offset relative to the intact precursor mass. In both cases, the characteristic mass (or mass difference) is fixed and is looked upon in all acquired MS/MS spectra, irrespectively of precursor m/z. Several CID pathways, however, produce combinations of fragments, whose masses differ in species of the same class, yet they are accountable by the same arithmetical equation. It is hardly possible to use these masses in conventional PIS or NLS, because taken alone, they are not specific for the lipid class. However, these combinations of fragment masses could be recognized in MS/MS spectra, if several selection criteria bundled by Boolean logical operations, are simultaneously applied. These complex data interrogation routines, termed Boolean scans, are performed in parallel with emulating conventional PIS and NLS and reveal important structural details regarding fragmented lipid precursors.

Figure 10.2A presents a part of the TOF MS survey scan spectrum of bovine heart polar lipid extract and two representative MS/MS spectra, acquired from precursors with m/z 792.56 (Fig. 10.2B) and m/z 752.57 (Fig. 10.2C) in the course of DDA-driven profiling. LipidInspector determined that both precursors belong to phosphatidylethanolamine (PE) class, because intense peaks of the neutral loss $\Delta m/z$ 141.02 products were observed at m/z 651.540 and m/z 611.544. Accurate mass of the latter fragment indicated that it was an ether lipid, although it was not possible to tell whether it belonged to the 1–O–alkyl or 1–O–alk–1′–enyl (plasmalogen) class.

In Positive ion mode plasmalogen species, contrary to 1–O–alkyl PE–species, produce a pair of characteristic fragments that retain moieties of the fatty acid and fatty alcohol, respectively, and the sum of their m/z equals M+2, where M stands for the neutral mass of a plasmalogen lipid (Zemski Berry and Murphy, 2004). Therefore, LipidInspector was set to identify in each spectrum, besides other fragment peaks, pairs of fragments $(m/z)_1$ and $(m/z)_2$ with intensities $I(m/z)_1$ and $I(m/z)_2$ that meet the following Boolean–type expression:

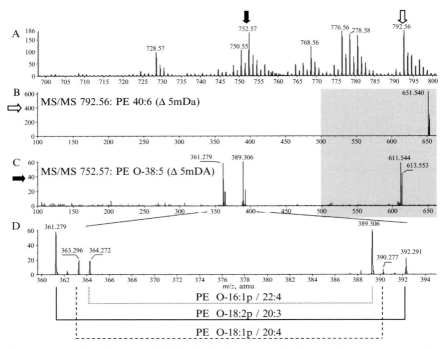

Figure 10.2 Identification of PE 1-O-alk-1′-enyl species by the plasmalogen-specific Boolean scan. Panel A: TOF MS survey spectrum of the polar extract of bovine heart. Although all plausible precursors were fragmented during DDA-driven MS/MS experiments, for the sake of clarity only two representative examples are presented in Panels B and C. The assignment of detectable fragments is explained in the text. Panel D: Zoom of the range of m/z 350 to 400 from the spectrum in Panel C. Using the plasmalogen-specific Boolean scan, LipidInspector identified three individual molecular species of plasmalogens, having exactly the same sum composition.

$$\{250 < [(m/z)_1; (m/z)_2] < 450\} \text{ AND } \{(m/z)_1 + (m/z)_2 = M + 2\}$$
$$\text{AND } \{I(m/z)_1/I(m/z)_2 = 0.3 \pm 0.1\} \text{ AND}$$
$$\{\text{peak@M} - 140\} = \text{TRUE}$$

The above expression effectively requested that a pair of candidate fragments should simultaneously meet the following four criteria:

1. Both fragments are within some practical range of m/z (in this case, above m/z 250 and below m/z 450).
2. The sum of their m/z equals M + 2.
3. The ratio of their intensities equals approximately 0.3 (with *ca* 30% tolerance).
4. Finally, the same spectrum contains a fragment corresponding to the neutral loss of PE head group.

We additionally requested (although, for presentation clarity, this is not included in the Boolean expression above) that nominal m/z of one of the two fragments should be even (this mass corresponds to the loss of the neutral fragment containing a fatty acid ester), whereas nominal m/z of another fragment should be odd (it is produced by the loss of the neutral fragment containing the PE head group and a vinyl alcohol; Zemski Berry and Murphy, 2004). If such a fragment pair was identified, LipidInspector annotated the precursor as a plasmalogen and reported the intensities of corresponding fragment peaks for further plasmalogen-specific quantification. We note that the analysis was not affected by co-fragmentation of several plasmalogen species (see Fig. 10.2), or ether PEs or lipids from other classes.

In the current version of LipidInspector, these criteria are bundled under "plasmalogen specific scan," which could be optionally selected by the user, along with other precursor and neutral loss scans, in a check-box menu. However, Boolean scans are generic and, in principle, could take advantage of a variety of CID mechanisms. Therefore, in the future, logical expressions should be programmed such that educated users could link several arithmetic equations and inequalities by various logical operations within a generic meta-language framework.

5. COMPREHENSIVE CHARACTERIZATION OF TOTAL LIPID EXTRACTS BY DDA-DRIVEN PROFILING

DDA-driven profiling characterizes the molecular composition of multiple lipid classes in parallel, which makes it a valuable tool for lipidomics screens. To test its performance, we applied the method for the partial characterization of *Caenorhabditis elegans* lipidome. Total lipid extracts were directly infused into a QSTAR Pulsar *i* mass spectrometer using NanoMate HD ion source. Tandem mass spectra were acquired under DDA control and then interpreted by LipidInspector program. While screening MS/MS spectra, LipidInspector was set to emulate three lipid class–specific NLSs, one lipid class–specific PIS, and one Boolean plasmalogen-specific scan (Fig. 10.3). As anticipated, PEs and PCs accounted for the major peaks detected in the survey TOF spectrum (Fig. 10.3A,B, and C). Interestingly, the major plasmalogen species (PE O–18:1p/18:1) was identified at m/z 730.58, while at m/z 732.58 NLS for $\Delta m/z$ 141.02, together with the accurately determined m/z of the corresponding fragment, identified the ether lipid PE O–36:1. With the exception of another two plasmalogen species, other peaks detectable by the same NLS were diacyl PEs. Peaks at m/z 820.74 and 822.74 were ammonia adducts of TAG 48:2 and TAG 48:1,

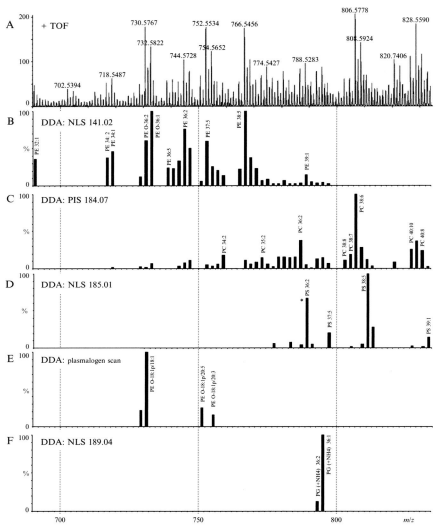

Figure 10.3 DDA-driven profiling of a total lipid extract from *Caenorhabditis elegans*. (A) Survey TOF MS spectrum. (B, C, D) Lipid class–specific profiles obtained by interpreting MS/MS spectra by LipidInspector program, which emulated : (B) neutral loss scan $\Delta m/z$ 141.02, specific for all species having a phosphatidylethanolamine head group (PEs, ether PEs, and plasmalogens); (C) precursor ion scan for the fragment with m/z 184.07, specific for PCs; (D) neutral loss scan for $\Delta m/z$ 185.01, specific for PSs; and (E) Boolean scan specific for plasmalogens. Details are presented in the text and in Fig. 10.2. (D) neutral loss scan $\Delta m/z$ 189.04, specific for PGs in the context of this experiment. The intensities of peaks of identified species were normalized to the total intensity of all species within the class. (With permission, from Schwudke, D., Oegema, J., Burton, L., Entchev, E., Hannich, J. T., Ejsing, C. S., Kurzchalia, T., and Shevchenko, A. (2006). Lipid profiling by multiple precursor and neutral loss scanning driven by the data-dependent acquisition. *Anal. Chem.* **78,** 585–595.)

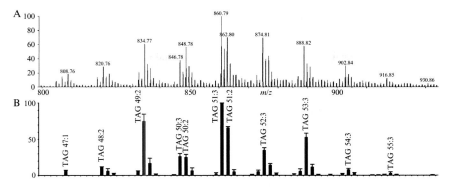

Figure 10.4 The composition of TAGs in *Caenorhabditis elegans*. (A) TOF survey spectrum of TAGs extracted from a TLC band. Note that peaks represent ammonium adducts of TAGs. (B) TAG profile produced by the emulation of 69 parallel neutral loss scans by LipidInspector. TAGs are annotated by their sum compositions and aligned with the spectrum in (A). To calculate the relative abundance of TAG peaks, the intensities of fatty acid loss fragments in each MS/MS spectrum were summed up. (With permission, from Schwudke, D., Oegema, J., Burton, L., Entchev, E., Hannich, J. T., Ejsing, C. S., Kurzchalia, T., and Shevchenko, A. (2006). Lipid profiling by multiple precursor and neutral loss scanning driven by the data-dependent acquisition. *Anal. Chem.* 78, 585–595.)

although the bulk of extractable TAGs were detected at the higher m/z range and are presented separately in Fig. 10.4.

Altogether, DDA-driven identification recognized 90 glycerophospholipids with unique sum compositions and provided their relative (within a given class) quantification in a single experiment. TAGs were analyzed in a separate experiment from the extract of the TLC band. While deciphering MS/MS spectra, LipidInspector emulated 69 scans specific for the neutral loss of fatty acid moieties from molecular adducts of TAGs with ammonium cations. We assumed that fatty acids comprised 9 to 22 carbon atoms and 0 to 6 double bonds. The identified TAGs were annotated with sum formulas and their relative abundance was determined (see Fig. 10.4). However, in all cases the emulated multiple neutral loss scans identified fatty acid moieties that were present in the fragmented precursors. Assuming that the abundance of neutral loss products does not strongly depend on fatty acid moieties and their position at the glycerol backbone, it was also possible to estimate the relative abundance of species with individual fatty acid compositions or, at least, identify major components in the mixture of isobaric precursors (Schwudke *et al.*, 2006).

This and other experiments demonstrated that complex lipid mixtures are fully amenable to DDA-driven profiling, which provides more specific information on the identity of lipid species compared to conventional precursor and neutral loss scans. In perspective, DDA profiling guided by comprehensive inclusion lists of plausible precursors are expected to produce an ultimately complete lipidomic data set when all fragment ions yielded from all possible precursors become detectable. Once such a data set is acquired, the software can re-process it in several user-specified ways with no recourse to repeating "wet" biology or MS/MS acquisition experiments.

6. SELECTED REACTION MONITORING IN QUANTIFICATION OF CHOLESTEROL AND CHOLESTERYL ESTERS

Compared to QqTOF instruments, triple quadrupole machines QqQ lack mass resolution and scan speed of the TOF analyzer, and therefore acquiring full MS/MS spectra from all possible precursor masses is impractical. However, the DDA approach specifically targets precursors of interest and use selected reaction monitoring (SRM) instead of conventional PIS for their quantification. Quantitative analysis of free cholesterol and cholesteryl ester in SRM mode serves as a good example. Cholesteryl esters (CE) are readily ionized as ammonium adducts, which produce a fragment ion of m/z 369.3 upon their CID. To determine free cholesterol content, it was first acetylated and ammonium adduct of the cholesterol acetate subjected to MS/MS analysis. Quantification was achieved in SRM mode using spiked synthetic internal standards—that is, D_7-cholesterol and CE17:0 and CE22:0—as reference compounds (Fig. 10.5) (Liebisch *et al.*, 2006).

SRM outperforms conventional or DDA-driven, lipid class–specific scans in the analysis of a small number of very low abundant precursors, especially if only a single fragment ion should be monitored and there is no need in acquiring full MS/MS spectra. For example, a human LDL fraction was analyzed in six replicates by PIS m/z 369 and by SRM targeting only CE precursors identified by survey PIS (see Fig. 10.5). PIS was performed within the range of m/z 600 to 760 with a scan speed of 200 amu/s. Assuming that peak width at half maximum was close to 0.7 amu, the effective scan time was 3.5 ms per analyzed species. In the direct flow injection analysis performed as described above, 50 scans were averaged to produce the spectrum for reliable quantification. However, in this analysis, each precursor of interested was analyzed for only 175 ms. In contrast, within the same time frame, SRM analysis of CEs, including acetyl ester of free cholesterol (see Fig. 10.5), was performed with a dwell time of 70 ms and 18 SRMs were averaged. Hence, each precursor was analyzed for 1260 ms. This difference in scan time enhanced the quantification accuracy. Major species, such as CE 18:2, 18:1, or 16:0, were quantified with CVs below 5% in PIS mode, compared to ∼2% using SRM. However, low abundant species like CE 20:3 or CE 24:1 were quantified with CVs of only ∼15% in PIS, whereas less than ∼5% was achieved by SRM. In summary,

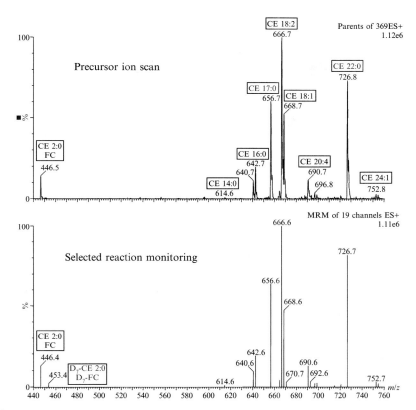

Figure 10.5 PIS *m/z* 369 and SRM for CE and FC quantification. Free cholesterol was acetylated before analysis. The upper panel shows a PIS *m/z* 369.3 of a human LDL fraction in positive ion mode specific for CE. The lower panel shows the spectrum of the same sample reconstructed from SRM data. Note the internal standard for quantifying free cholesterol D₇–CE 2:0 was detected using different transition (*m/z* 453.4→376.3). CE, cholesteryl ester; FC, free cholesterol. (See color insert.)

for a moderate number of lipid precursors, SRM on triple quadrupole instruments produced significantly higher accuracy, compared to experiments in scanning mode. SRM driven by the inclusion list of rapid PIS used as a survey scan is especially beneficial for low abundance species and samples with known lipid species composition.

ACKNOWLEDGMENTS

We are grateful to colleagues at the Shevchenko laboratory for useful discussions on the lipidomic strategies; Igor Chernushevich and Lyle Burton (MDS Sciex) for expert advice on quadrupole time-of-flight mass spectrometry; Jeff Oegema (Scionics Computer Innovations

GmbH) for collaboration on LipidInspector software; and Judith Nicholls (MPI of Molecular
Cell Biology and Genetics) for critical reading of the manuscript. Funded in part by the
Deutsche Forschungsgemeinschaft (SFB/TR 13 grant) to AS (project D1) and GS (project
A3), and to GL (Li 923/2.1/2).

REFERENCES

Bligh, E. G., and Dyer, W. J. (1959). A rapid method of total lipid extraction and purification. *Can. J. Biochem. Physiol.* **37,** 911–917.

Brugger, B., Erben, G., Sandhoff, R., Wieland, F. T., and Lehmann, W. D. (1997).
Quantitative analysis of biological membrane lipids at the low picomole level by
nanoelectrospray ionization tandem mass spectrometry. *Proc. Natl. Acad. Sci. USA* **94,**
2339–2344.

Chernushevich, I., Loboda, A., and Thomson, B. (2001). An introduction to quadrupole
time-of-flight mass spectrometry. *J. Mass Spectrom.* **36,** 849–865.

Ejsing, C. S., Duchoslav, E., Sampaio, J., Simons, K., Bonner, R., Thiele, C., Ekroos, K.,
and Shevchenko, A. (2006a). Automated identification and quantification of
glycerophospholipid molecular species by multiple precursor ion scanning. *Anal. Chem.*
78, 6202–6214.

Ejsing, C. S., Moehring, T., Bahr, U., Duchoslav, E., Karas, M., Simons, K., and
Shevchenko, A. (2006b). Collision-induced dissociation pathways of yeast sphingolipids
and their molecular profiling in total lipid extracts: A study by quadrupole TOF and
linear ion trap–orbitrap mass spectrometry. *J. Mass Spectrom.* **41,** 372–389.

Ekroos, K., Chernushevich, I. V., Simons, K., and Shevchenko, A. (2002). Quantitative
profiling of phospholipids by multiple precursor ion scanning on a hybrid quadrupole
time-of-flight mass spectrometer. *Anal. Chem.* **74,** 941–949.

Ekroos, K., Ejsing, C. S., Bahr, U., Karas, M., Simons, K., and Shevchenko, A. (2003).
Charting molecular composition of phosphatidylcholines by fatty acid scanning and ion
trap MS3 fragmentation. *J. Lipid Res.* **44,** 2181–2192.

Ekroos, K., and Shevchenko, A. (2002). Simple two-point calibration of hybrid quadrupole
time-of-flight instruments using a synthetic lipid standard. *Rapid Commun. Mass Spectrom.*
16, 1254–1255.

Han, X., and Gross, R. W. (2003). Global analyses of cellular lipidomes directly from crude
extracts of biological samples by ESI mass spectrometry: A bridge to lipidomics. *J. Lipid
Res.* **44,** 1071–1079.

Han, X., and Gross, R. W. (2005). Shotgun lipidomics: Electrospray ionization mass
spectrometric analysis and quantitation of cellular lipidomes directly from crude extracts
of biological samples. *Mass Spectrom. Rev.* **24,** 367–412.

Han, X., Yang, K., Yang, J., Fikes, K. N., Cheng, H., and Gross, R. W. (2006). Factors
influencing the electrospray intrasource separation and selective ionization of glycerophospholipids. *J. Am. Soc. Mass Spectrom.* **17,** 264–274.

Liebisch, G., Binder, M., Schifferer, R., Langmann, T., Schulz, B., and Schmitz, G. (2006).
High throughput quantification of cholesterol and cholesteryl ester by electrospray
ionization tandem mass spectrometry (ESI-MS/MS). *Biochim. Biophys. Acta* **1761,**
121–128.

Liebisch, G., Lieser, B., Rathenberg, J., Drobnik, W., and Schmitz, G. (2004). High-
throughput quantification of phosphatidylcholine and sphingomyelin by electrospray
ionization tandem mass spectrometry coupled with isotope correction algorithm.
Biochim. Biophys. Acta **1686,** 108–117.

Makarov, A., Denisov, E., Kholomeev, A., Balschun, W., Lange, O., Strupat, K., and Horning, S. (2006). Performance evaluation of a hybrid linear ion trap/orbitrap mass spectrometer. *Anal. Chem.* **78,** 2113–2120.

Pulfer, M., and Murphy, R. C. (2003). Electrospray mass spectrometry of phospholipids. *Mass Spectrom. Rev.* **22,** 332–364.

Schwudke, D., Oegema, J., Burton, L., Entchev, E., Hannich, J. T., Ejsing, C. S., Kurzchalia, T., and Shevchenko, A. (2006). Lipid profiling by multiple precursor and neutral loss scanning driven by the data-dependent acquisition. *Anal. Chem.* **78,** 585–595.

Schwudke, D., Hannich, J. T., Surendranath, V., Grimard, V., Moehring, T., Burton, L., Kurzchalia, T., and Shevchenko, A. (2007). Top-down lipidomic screens by multivariate analysis of high-resolution survey mass spectra. *Anal. Chem.* **79,** 4083–4093.

Syka, J. E., Marto, J. A., Bai, D. L., Horning, S., Senko, M. W., Schwartz, J. C., Ueberheide, B., Garcia, B., Busby, S., Muratore, T., Shabanowitz, J., and Hunt, D. F. (2004). Novel linear quadrupole ion trap/FT mass spectrometer: Performance characterization and use in the comparative analysis of histone H3 post-translational modifications. *J. Proteome Res.* **3,** 621–626.

Wenk, M. R. (2005). The emerging field of lipidomics. *Nat. Rev. Drug Discov.* **4,** 594–610.

Zemski Berry, K. A., and Murphy, R. C. (2004). Electrospray ionization tandem mass spectrometry of glycerophosphoethanolamine plasmalogen phospholipids. *J. Am. Soc. Mass Spectrom.* **15,** 1499–1508.

CHAPTER ELEVEN

IDENTIFICATION OF INTACT LIPID PEROXIDES BY AG$^+$ COORDINATION ION-SPRAY MASS SPECTROMETRY (CIS-MS)

Huiyong Yin *and* Ned A. Porter

Contents

Abstract

Free radical–induced autoxidation of lipids containing polyunsaturated fatty acids (PUFA) has been implicated in numerous human diseases including atherosclerosis, neurodegenerative diseases, and cancer. Autoxidation of PUFAs generates hydroperoxides as primary oxidation products and further oxidation leads to cyclic peroxides as secondary oxidation products. It is

Departments of Chemistry and Medicine, Division of Clinical Pharmacology, Center in Molecular Toxicology, Vanderbilt Institute of Chemical Biology, Vanderbilt University, Nashville, Tennessee

Methods in Enzymology, Volume 433
ISSN 0076-6879, DOI: 10.1016/S0076-6879(07)33011-5

challenging to identify these peroxides by conventional electro-spray ionization (ESI) mass spectrometry (MS) method because of their thermal and chemical instability under most analytical conditions. Ag^+ coordination ion-spray CIS-MS has proven to be a powerful tool to analyze these intact lipid peroxides. Ag^+ preferentially complexes with the double bonds and induces characteristic fragmentation in the gas phase. Monocyclic peroxides, bicyclic endoperoxides, serial cyclic peroxides, and dioxolane-isoprostane peroxides have been identified from the oxidation of cholesteryl arachidonate and phospholipids containing arachidonate. This technique has been widely used for structural identification but it is difficult to make it a quantitative tool because of the formation of multiple silver adducts.

1. INTRODUCTION

Lipids are essential components of cellular membranes that maintain the biological function of mammalian cells. Lipids and the metabolites derived from lipids also play an important role in the control of major cellular activities, such as signal transduction (Roberts and Morrow, 2002). Polyunsaturated fatty acids (PUFA) are primary targets for attack by reactive oxygen species or free radicals, and the oxidation of these lipids in humans is frequently associated with various diseases, such as atherosclerosis, cancer, diabetes, chronic inflammatory bowel disease, rheumatoid arthritis, and neurodegenerative disorders including Alzheimer and Parkinson diseases (Montine and Morrow, 2005; Porter et al., 1995). Thus, lipid peroxidation, or reaction of a lipid with molecular oxygen, has been an intensive research area for decades (Porter, 1986; Yin and Porter, 2005). Tremendous research efforts have been made to understand the mechanisms of lipid peroxidation and prevent the deleterious effect of lipid autoxidation. It is generally believed that free radical reactions have been involved in the oxidation of PUFAs to generate hydroperoxides and other highly oxidized products. The peroxides can also degrade to give highly reactive electrophiles, such as malonaldehyde (MDA), acrolein and 4-hydroxynonenal (4-HNE), which may covalently modify protein and DNA (Lee et al., 2001; Poli and Schaur, 2000; Shen and Ong, 2000; Uchida, 2000; Uchida et al., 1998).

The free radical mechanisms which lead to the oxidation of PUFA-containing lipids have been well studied. An example of arachidonate oxidation—cholesterol esters or phospholipids—is summarized in Fig. 11.1. Under free radical conditions, hydrogen atom abstraction from one of the three bis–allylic sites on carbon 7, 10, and 13 of arachidonate generates a delocalized pentadienyl radical such as 1b. Molecular oxygen addition to this carbon-centered radical gives rise to a peroxyl radical 1c, which can lead to the formation of primary hydroperoxides 1d (11-hydroperoxyleicosatetraenoic acid,

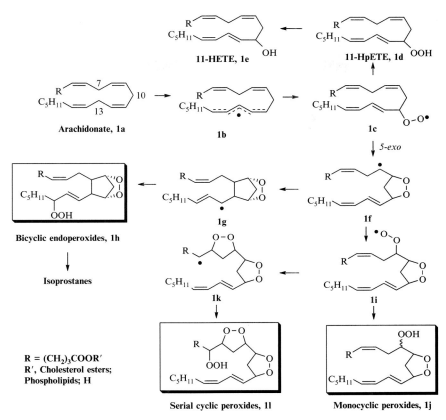

Figure 11.1 Free radical mechanisms that lead to the formation of primary and secondary oxidation products from arachidonate (cholesterol esters and phospholipids).

11-HpETE) and 11-HETE, 1e. Alternatively, peroxyl radical 1c can cyclize to form radical 1f. Bicyclic endoperoxides, 1h, can be formed from 1f via 1g. The bicyclic endoperoxides and subsequent reduced or rearranged products are isomers of prostaglandins derived from cyclooxygenases (COX-1 or COX-2). These compounds are formed *in vivo* in significant amounts under conditions of oxidative stress and quantification of them has evolved as the "gold standard" to assess oxidative stress status *in vivo* (Morrow, 2005; Morrow *et al.*, 1990). Molecular oxygen addition to radical 1f generates peroxyl radical 1i, which undergoes H-abstraction to give rise to monocyclic peroxides 1j. More complex peroxides, such as the serial cyclic compound 1l, can be formed from further cyclization of peroxyl radical 1i. Overall, oxidation of arachidonate produces a complex array of hydroperoxides and cyclic peroxides. Identification of these unstable peroxides poses a big challenge to the lipid research community.

2. ANALYSIS OF LIPID OXIDATION PRODUCTS

Analysis of lipid peroxidation products has been an extremely active area in the research field of lipid oxidation for decades (Moore and Roberts, 1998). Various methods have been developed to assess lipid peroxidation. Measurement of conjugated dienes with the UV absorption at 234 nm has been widely used to assess lipid peroxidation. However, the application of this method has been limited because of the complication of other compounds having the same absorption. Light emission during hydroperoxide-induced oxidation of isoluminol has also been employed as a chemiluminescence assay for lipid peroxidation (Miyazawa *et al.*, 1994). This sensitive method has been used to analyze plasma samples—levels of hydroperoxides from cholesterol esters and phospholipids—when combined with an HPLC separation technique. Iodometric assays are based on the reaction of lipid hydroperoxides with iodide to form iodine or triiodide, which can be measured by a spectrophotometric assay, an anaerobic assay, or a cadmium-based assay. An immunohistochemical technique has been used to raise antibodies against specific lipid peroxidation products, such as 4-HNE (Paradis *et al.*, 1997). The thiobarbituric acid reactive substance (TBARS) assay is one of the oldest and most frequently used methods to assess lipid peroxidation (Kikugawa, 1997). The test is usually performed by heating a lipid peroxidation mixture and TBA in an acidic medium to form a red pigment which shows an absorption maximum at 532 nm (with molecular extinction coefficient of 1.56×10^5) and fluorescence at 532 nm (excitation) and 553 nm (emission). The ferrous oxidation of xylenol (FOX) assay was developed by Wolff and coworkers to analyze the hydroperoxides formed from lipid peroxidation (Jiang *et al.*, 1991; Nourooz-Zadeh *et al.*, 1994). This method is based on the fact that hydroperoxides oxidize ferrous (iron II) to ferric (iron III), and the resulting ferric ion can bind to xylenol orange to produce a colored complex with a strong absorbance at 560 nm (ε $4.3 \times 10^4 \, M^{-1} cm^{-1}$ for H_2O_2 and t-butyl hydroperoxide). Both the TBARS and Fox assays are nonspecific methods and highly oxidized peroxides can give a positive response even after triphenylphosphine reduction (Yin and Porter, 2003).

3. Ag$^+$ COORDINATION ION-SPRAY MS

Although the abovementioned methods to analyze lipid oxidation products have been successfully applied to study *in vitro* lipid oxidation, identification of more complicated intact peroxides has been limited by their inherent thermal and chemical instability, and conventional characterization methods have proven difficult when applied to these compounds (Yin *et al.*, 2000, 2001). Mass spectrometry methods have the advantages of identifying the structure of these compounds and quantifying the amounts of different oxidation products. Gas chromatography (GC)-MS has been used in the analysis of lipid autoxidation products after derivatization (methyl esters or PFB esters and trimethylsilyl [TMS] ethers). To do this the peroxides formed during lipid peroxidation must be reduced to alcohols before being derivatized to PFB or methyl esters. Electron impact (EI)-MS of these derivatives gives specific fragmentation information, such as α-cleavage, which can be used to identify the position of a hydroxyl group in the lipid chain. In addition to EI-MS, the alcohols also can be analyzed under chemical ionization (CI)-MS conditions. In this protocol, the alcohols are converted to the PFB esters for analysis under negative-ion electron capture conditions (Guido *et al.*, 1993; Morrow *et al.*, 1999).

Direct analysis of lipid peroxidation products without prior derivatization would be of significant importance because multiple derivatization steps are time-consuming and allow plenty of opportunities for sample loss or *ex vivo* lipid oxidation during the sample workups. There are a few successful examples in the direct analysis of lipid hydroperoxides. Reverse-phase HPLC coupled with thermo-spray ionization has been used to identify lipid hydroperoxides and hydroperoxyl fatty acids esterified to phospholipids. Murphy *et al.* (2001) applied electro-spray ionization MS to the direct analysis of lipid hydroperoxides and other oxidation products (Harrison *et al.*, 2000; MacMillian and Murphy, 1995; Murphy *et al.*, 2001). The structure identification for these intact peroxides as illustrated in Fig. 11.1 has proven to be difficult. ESI in the negative ion mode allowed the generation of abundant carboxylate anions from free-acid hydroperoxides. Using tandem MS techniques, distinctive fragmentation patterns were obtained, which makes structural elucidation possible.

Analysis of intact cholesteryl ester or phospholipid hydroperoxides derived from low-density lipoprotein (LDL) is difficult to undertake by conventional MS methods. Obviously, GC-MS would not be suitable because these thermally labile compounds cannot survive GC conditions. Electro-spray ionization (ESI) and APCI are widely used soft ionization modes and may be coupled efficiently to powerful separation techniques such as HPLC for the characterization of complex mixtures. Nonetheless, cholesteryl esters are highly lipophilic and lack ionization sites with sufficient proton affinity that are required for ESI and APCI applications (Blakely and Adams, 1978; Dole *et al.*, 1968; Gaskell, 1997).

In 1999, Bayer *et al.* reported the use of coordinating reagents to ionize highly lipophilic compounds, such as terpenes, sugars, aromatics, and vitamins (Bayer *et al.*, 1999). A variety of coordinating agents were used to coordinate with the analytes, and the resulting complexes were made suitable for MS analysis through a standard ESI-MS interface. This technique is termed coordination ion-spray mass spectrometry (CIS-MS). Silver ion has been used extensively for CIS-MS applications because it is a soft Lewis acid, and thus is prone to coordinate with double bonds or aromatics. As an example, the use of silver impregnated silica gel has been routinely employed to significantly improve the separation of unsaturated lipids by chromatography (Morris, 1966).

Silver ion adducts show characteristic complexes of $[M + Ag^{107}]^+$ and $[M + Ag^{109}]^+$ doublets in the mass spectrum, because the natural isotopic abundance of silver is 52:48. CIS-MS can also be coupled with online LC separations for analysis of complex mixtures of products. Ag^+ CIS-MS has also been used to characterize lipophilic tocopherols and carotenoids (Rentel et al., 1998; Zhang and Brodbelt, 2005).

4. IDENTIFICATION OF LIPID OXIDATION PRODUCTS OF CHOLESTROL ESTERS BY LC-AG+ CIS-MS

Cholesterol esters such as cholesteryl linoleate (C18:2) and cholesteryl arachidonate are major nonpolar lipids in the core of low-density lipoprotein particles. According to the free radical mechanisms listed in Fig. 11.1, four isomeric hydroperoxides can be generated in the oxidation of cholesteryl linoleate including 9-t,c, 9-t,t-, 13-t,c, and 13-t,t (t, trans; c, cis) hydroperoxides. In the case of cholesterol arachidonate, six conjugated hydroperoxides can be formed under free radical conditions. These primary hydroperoxides can be further oxidized to give rise to array of cyclic peroxides, such as monocyclic peroxides, serial cyclic peroxides, bicyclic endoperoxides, and dioxolane-endoperoxides (Yin and Porter, 2005). Analysis of these intact peroxides from cholesterol esters is more difficult because they lack protonation or deprotonation sites and they will not survive basic hydrolysis. Ag^+ CIS-MS has the advantage of generating charged species under mild electro-spray conditions, and Ag^+ coordination generates rich fragmentation in the gas phase, which can be used to identify these novel structures (Havrilla et al., 2000; Yin et al., 2002).

4.1. Generation of oxidation products from cholesterol arachidonate *in vitro*

Cholesteryl arachidonate (0.400 g, 0.594 mmol) was dissolved in dry benzene (4.0 ml) for a final lipid concentration of 0.15 M. Oxidation was initiated by the addition of a free radical initiator such as di-*tert*-butyl hyponitrite (DTBN; ~3 mg). The reaction was placed into a 37° oil bath under air for 48 h. The temperature was controlled at 37.0 ± 0.2° using an

I^{2}R Thermo-watch ML6-1000SS. The reaction was quenched by the addition of butylated hydroxytoluene (BHT) ($\sim$2 mg). A portion of the reaction mixture (1 ml) was removed and concentrated to dryness. The residue was diluted with CHCl$_3$ (100 μl), and applied to a Burdick & Jackson Inert SPE System (silica, 500 mg) preconditioned with hexane (10 ml) and equilibrated with 1.5% methyl *tert*-butyl ether (MTBE) in hexane (10 ml). The unoxidized cholesteryl arachidonate was eluted with 1.5% MTBE in hexanes (3 × 5 ml). The oxidized cholesteryl arachidonate material was eluted from the column with 5% methanol in MTBE (3 × 5 ml). The oxidized fractions were combined and concentrated to yield about 61.2 mg of oxidized material. A stock solution of the oxidized material was prepared in benzene (2 ml) for a concentration of 30.6 mg/ml. Subsequent dilutions of this stock solution were used for mass spectrometric analysis. Analytical HPLC analysis (single Si column, 1.0% IPA in hexane) indicated no difference in the product distribution of polar products after removal of the unoxidized lipid.

4.2. Analysis of oxidation products from cholesteryl arachidonate by Ag$^+$ CIS-MS

Analytical HPLC was carried out using a Waters Model 600E pump with a Waters 996 Photodiode array detector. Millenium32 chromatography software (Waters Corp., Milford, MA) was used to control the 996 and to collect and process data. Cyclic peroxide analysis by analytical HPLC utilized a single Beckman Ultrasphere 5 μ (4.6 mm × 25 cm) silica column. A flow rate of 1 ml/min was used for analytical normal phase (NP) HPLC. Preparative NP HPLC was performed using a Dynamax-60 Å 8 μ (83–121-C) silica column (21.4 mm × 25 cm) with a flow rate of 10 ml/min. A single Beckman Ultrasphere 5 μ (2.0 mm × 25 cm) silica column was used for analysis of cyclic peroxides, and two Beckman Ultrasphere 5 μ (2.0 mm × 25 cm) silica columns for acyclic hydroperoxide analysis. Flow rate for the LC-MS separations and analysis was 150 μl/min.

LC-CIS-MS was performed using a Hewlett Packard 1090 series pump connected to a Finnigan TSQ-7000 or TSQ Quantum 1.0 SR 1 (San Jose, CA) triple-quadrupole mass spectrometer equipped with a standard electrospray ionization source. The source was outfitted with a 100-μm i.d. deactivated fused Si capillary. Data acquisition and evaluation were conducted using ICIS EXECUTIVE INST, version 8.3.2, and TSQ 7000 software INST, version 8.3, or Xcalibur, version 2.0, on quantum instrument. Nitrogen gas served both as sheath gas and auxiliary gas; argon served as the collision gas. The electro-spray needle was maintained at 4.6 kV, and the heated capillary temperature was 200°. The tube lens potential and capillary voltage were optimized to maximize ion current for electro-spray, with the optimal determined to be 80 V and 20 V, respectively, for cholesteryl ester analysis.

Positive ions were detected scanning from 100 to 1000 amu with a scan duration of 2 s. Profile data were recorded for 1 min and averaged for analysis. For collision-induced dissociation (CID) experiments, the collision gas pressure was set from 2.30 to 2.56 mTorr on TSQ 7000 and 1.5 mTorr on TSQ Quantum. To obtain fragmentation information of each compound, the dependence of offset voltage and relative ion current (RIC) was studied. The collision energy offset varied from 10 to 40 eV depending on the compound analyzed.

Samples were introduced either by direct liquid infusion or by normal phase HPLC. For direct liquid injection, stock solutions of the lipids (100 ng/μl in 1% isopanol [IPA] in hexane) were prepared and mixed 1:1 with AgBF$_4$, silver tetrafluoroborate (51.4 ng/μl in IPA). Samples were introduced to the ESI source by syringe pump at a rate of 10 μl/min. For HPLC sample introduction, a Hewlett-Packard 1090 HPLC system was used. The auxiliary gas flow rate to the ESI interface was increased to between 5 to 10 units to assist in desolvation of the samples. For cholesteryl linoleate and cholesteryl arachidonate hydroperoxide analysis, normal–phase HPLC sample introduction was carried out using two tandem Beckman Ultrasphere narrow-bore 5 μ silica columns (2.0 mm × 25 cm) operated in isocratic mode with 0.35% isopropanol in hexanes. For analysis of cyclic peroxide mixtures, sample introduction was carried out using a single Beckman Ultrasphere narrow-bore 5 μ silica column (2.0 mm × 25 cm) operated in isocratic mode with 1.0% isopropanol in hexanes. The flow rate for both modes of chromatography was 150 μl/min. Column effluent was passed through an Applied Biosystems 785A Programmable Absorbance UV detector with detection at $\lambda = 234$ nm. An Upchurch high-pressure mixing tee was connected next in series for the post-column addition of the silver salts. The silver tertafluoroborate (AgBF$_4$) solution (0.25 mM in isopropanol) was added via a Harvard Apparatus (Cambridge, MA) syringe pump at a flow rate of 75 μl/min. A long section of PEEK tubing (1.04 m, 0.25 mm i.d.) allowed time for the complexation of the silver to the lipid while delivering effluent to the mass spectrometer. A Rheodyne 7725 injector was fitted with a 100-μl PEEK loop for 20- to 50-μl sample injections.

4.3. Results

Ag$^+$ CIS-MS has been successfully applied to locate the hydroperoxide functional groups in the oxidation products of cholesterol esters based on the characteristic fragmentation induced by silver ion coordination (Havrilla et al., 2000). Characterization of 5-HpETE Ch is taken as an example. The Q1 scan of 5-HPETE Ch and the CID spectrum of m/z 811 are shown in Fig. 11.2. The characteristic doublet peaks of m/z 811/813 correspond to the HpETE Ch and Ag$^+$ adduct. The structural information of the parent ion is obtained by a CID experiment which is carried out by selecting

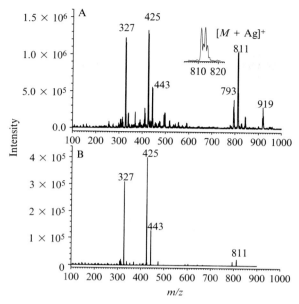

Figure 11.2 5-HpETE Ch analyzed by Ag⁺ CIS-MS. (A) Parent ion q1 scan. (B) Collision-induced dissociation of m/z 811. Ch, cholesterol.

m/z 811 in the first quadrupole, fragmenting it in the second quadrupole and scanning all the resulting fragments in the third. The major fragments from the parent ion are summarized in Fig. 11.3A. Beside the loss of cholesterol and subsequent dehydration, there is a fragment with m/z 327 that is consistent with the structure of an aldehyde formed from fragmentation of 5-HpETE. The formation of the aldehyde can be understood based on well-established Hock fragmentation of hydroperoxides (Fig. 11.3B). It is clear from these data that Hock fragmentation can give regioisomeric information about the hydroperoxide functionality of the HpETEs Ch. Other hydroperoxide regiosiomers of cholesteryl arachidonates can be studied the same way as for 5-HPETE Ch and they show characteristic Hock fragments. Coupled with HPLC, selective reaction monitoring (SRM) is used to identify individual hydroperoxides in oxidation mixtures. SRM is performed in a triple-quadrupole tandem mass spectrometer by selecting the parent ion (m/z 811 for HpETEs Ch) in the first quadrupole, fragmenting it in the second quadrupole, and monitoring a specific fragment (Hock fragments for HpETEs Ch) resulting from the parent ion in the third. The SRM results of the six conjugated hydroperoxides of cholesteryl arachidonate are summarized in Fig. 11.4. 8-HpETE Ch and 9-HpETE Ch, 11-HpETE, and 12-HpETE Ch give the same Hock fragments. Therefore, it is impossible to distinguish these isomeric hydroperoxides

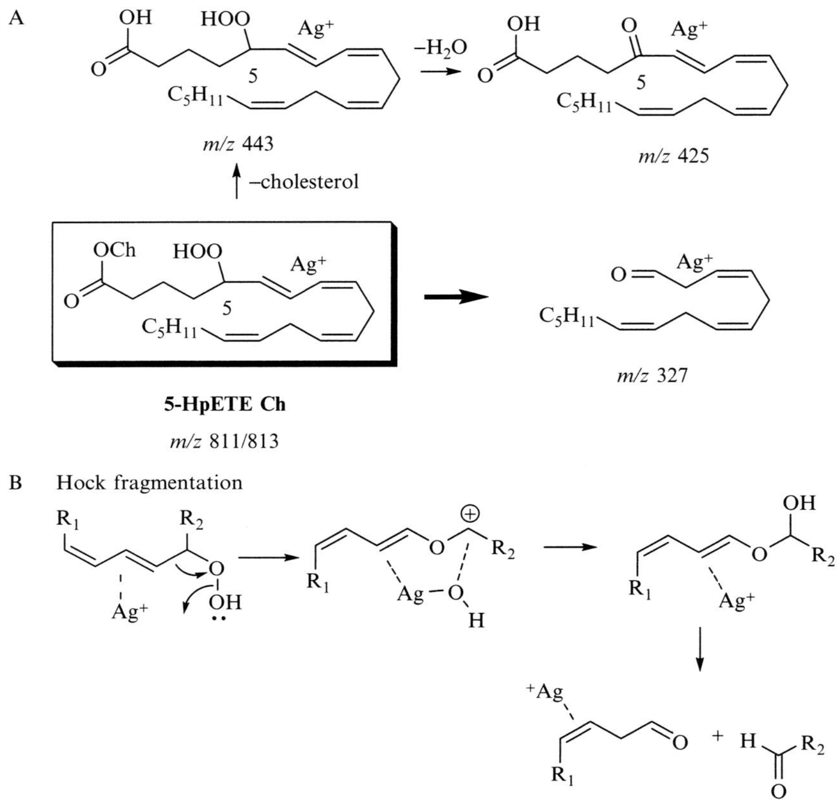

Figure 11.3 Pathways that lead to the fragmentation of silver adducts of 5-HpETE Ch in CIS-MS. (A) Gas phase fragmentation in collision-induced dissociation of m/z 811. (B) Hock cleavage of hydroperoxides induced by silver coordination.

based on Hock fragmentation alone. There is minor α-cleavage fragmentation in the CID experiment of HpETE Ch, and this fragmentation can be used to further distinguish the regioisomeric HpETEs Ch. It represented the first example of identification of intact cholesterol esters by the use of mass spectrometry.

Oxidation of polyunsaturated lipids such as arachidonate can produce many regioisomeric peroxyl radicals that lead to an extremely complex oxidation mixture (Fig. 11.1). Biological samples are even more challenging, because many types of polyunsaturated lipids are present, increasing the complexity of the lipid autoxidation profile. For *in vitro* studies, starting from a single isomeric hydroperoxide dramatically reduces the complexity of the autoxidation mixture. After identification of the primary oxidation products hydroperoxides, the individual hydroperoxides can be purified by

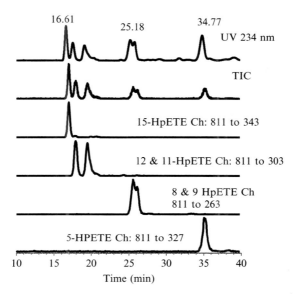

Figure 11.4 SRM results of conjugated HpETE Ch analyzed by Ag⁺ CIS-MS.

semipreparative HPLC. 15-HpETE Ch, for example, is used as a model compound for the analysis of the autoxidation products by silver ion LC-MS techniques. As already mentioned, monocyclic peroxides, serial cyclic peroxides, and bicyclic endoperoxides can be formed from the oxidation of 15-HpETE Ch. Dozens of stereoisomers within each class of peroxides are predicted to be formed because several stereogenic centers are formed in these processes (Yin et al., 2002).

The autoxidation mixture from reactions of 15-HpETE Ch was first analyzed by a direct liquid infusion (DLI) experiment, which was carried out by continuous infusion of a mixture of AgBF₄ solution and the autoxidation mixture into a triple-quadrupole tandem mass spectrometer operating in Q_1-scan mode. The DLI experiment clearly showed that masses consistent with all three classes of cyclic peroxides (bicyclic, monocyclic, and serial cyclic) are observed in the autoxidation mixture of 15-HpETE Ch in addition to the starting hydroperoxide (Fig. 11.5). CID experiments were carried out to further characterize these cyclic peroxides based on their characteristic fragmentation. An example of the CID spectrum of m/z 843 corresponding to a bicyclic endoperoxide and the possible fragmentation pathways are shown in Fig. 11.6. All the fragmentation reactions observed have well-established precedence in solution. The CID experiment unambiguously supports the presence of bicyclic endoperoxide structures in the product mixture. It is the first time that intact bicyclic endoperoxides have been identified by MS methods (Yin et al., 2002), which is of significance because these compounds are the precursors of

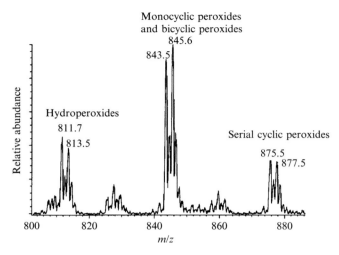

Figure 11.5   Direct liquid infusion results of oxidation mixture from autoxidation of 15-HpETE Ch.

the isoprostanes. These nonenzymatic products generated *in vivo* under oxidative stress have been implicated in a number of human diseases including atherosclerosis and neurodegenerative diseases (Morrow, 2005). These compounds possess potent biological activities, and quantification of these compounds represents the most accurate means to assess oxidative stress status *in vivo*. Furthermore, monocyclic peroxides and serial cyclic peroxides have also been identified by analogous CID experiments. SRM chromatograms of the oxidation products of 15-HPETE Ch revealed a number of diastereoisomers within each class of cyclic peroxides. The oxidation mixtures from 11-, 5-, and 9-HpETE Ch have also been studied by the same technique and similar cyclic peroxides have been identified. In addition, a novel class of cyclic peroxides has been identified from oxidation mixture of 12- and 8-HpETE (Yin *et al.*, 2004). These compounds are termed dioxolane-isoprostane peroxides because a dioxolane functional group and a bicyclic peroxide moiety are present in the same molecule. The novel structure of these compounds is further confirmed by other MS methods.

5. Identification of Intact Lipid Peroxides from Phospholipids by Ag+ CIS-MS

Glycerophospholipids such as phosphatidylcholines are major classes of lipids in cellular membranes and on the surface layer of LDLs. Linoleic acid and arachidonic acid are the most abundant unsaturated fatty acids esterified in LDL phospholipids. As can be predicted based on the free

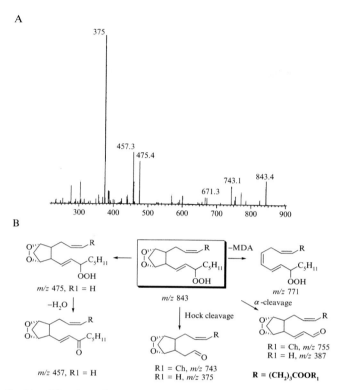

Figure 11.6 Identification of intact bicyclic endoperoxides cholesterol esters by Ag^+ CIS-MS. (A) Collision–induced dissociation spectrum. (B) Fragmentation pathway of bicyclic endoperoxide silver adduct with m/z 843.

radical mechanisms outlined in Fig. 11.1, hydroperoxides, monocyclic peroxides, bicyclic peroxides, and serial cyclic peroxides can be formed in the oxidation mixture of phospholipids. Even though MS has long been applied to study the oxidation products of phospholipids (Hall and Murphy, 1998; Pulfer and Murphy, 2003; Spickett et al., 1998), unambiguous structure identification is often difficult to achieve because of the predominant fragmentation resulting from the loss of head groups. The Ag^+ CIS-MS technique was applied to this problem, and it provides useful structural information (Milne and Porter, 2001).

5.1. Autoxidation of phospholipid 1-palmitoyl-2-linoleoyl-sn-glycero-3-phosphocholine (PLPC)

A solution of PLPC (25 mg, 0.033 mmol) in 2.5 ml of CH_2Cl_2 was evaporated to dryness under vacuum so that the mixture formed a thin layer on the inside of a 10-ml round–bottomed flask. The flask was then heated to

$37°$ and exposed to an atmosphere of dry air. After 24 h, the mixture was dissolved in benzene and BHT ($\sim$1 to 2 mg) was added to stop the reaction.

5.2. LC-CIS-MS analysis of oxidation products from PLPC

Samples were introduced either by direct liquid infusion (DLI) or HPLC. For DLI experiments, samples were introduced with a Harvard Apparatus (Cambridge, MA) syringe pump at a flow rate of 10 μl/min. For HPLC sample introduction, a Waters model 2690 Separation Model instrument was used. The HPLC was equipped with a Discovery octadecylsilane (ODS) column (4.6 × 250 mm, 5 μm; Supelco, Bellefonte, PA) and operated with a mobile phase of methanol/water (95:5, v/v) at a flow rate of 1 ml/min. A splitting tee after the column permitted 240 μl/min to be passed through an Applied Biosystems 785A programmable absorbance ultraviolet (UV) detector operating at 234 nm before entering the mass spectrometer. The remainder of the effluent was collected as waste. For these experiments, the hydroperoxides were isolated from the unoxidized phospholipid using analytical HPLC (methanol/water, 95:5, v/v, 1 ml/ min). The oxidized fraction for PLPC (13.5 to 15.5 min) was collected, concentrated, and analyzed by LC-CIS-MS. A stock solution of the oxidized lipid was prepared (1 to 1.25 mg/ml) and 25 to 50 μl of the solution was injected per analysis. Offset voltages for selected reaction monitoring (SRM) experiments, 28 to 33 eV, were determined by optimization in DLI experiments.

5.3. Results

Analysis of oxidized phospholipids such as PLPC by Ag^+ CIS-MS can be achieved by either post-column mixing of 0.5 mM of $AgBF_4$ in methanol with the HPLC effluent or by addition of $AgBF_4$ to the HPLC mobile phase to yield a 0.15-mM solution. When a mixture of oxidation products from PLPC was analyzed by Ag^+ CIS-MS, m/z 896/898 correspond to the silver adducts of PLPC hydroperoxides including 13-hydroperoxyoctadecyl phosphatidylcholine (13-HpODE PC) and 9-HpODE PC as in Fig. 11.7A and B. CID experiments on m/z 896 generates several fragments that result from cleavage of the head group choline, m/z 713 and 695, whereas Hock cleavage gives rise to a fragment of m/z 613 that is characteristic of 13-HpODE PC (Fig. 11.7c and d). The peak with m/z 753 is the Hock fragment of 9-HpODE PC. Therefore, silver ion adduction leads to the formation of characteristic fragments that can be utilized to locate the hydroperoxides in phospholipids. Oxidation of arachidonyl-containing phospholipids, such as 1-stearoyl-2-arachidonoyl-sn-glycero-3-phosphocholine (SAPC), generates a more complex mixture of oxidation products including hydroperoxides, monocyclic peroxides, bicyclic endoperoxides, and serial cyclic peroxides.

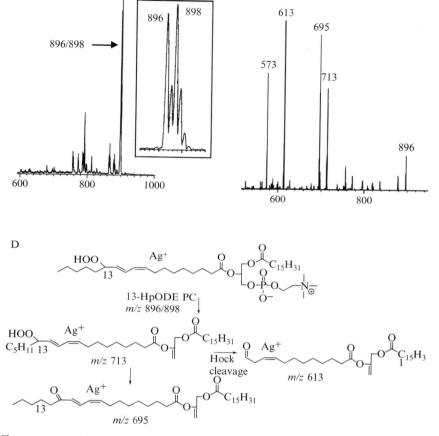

Figure 11.7 Ag$^+$ CIS-MS results of PLPC oxidation mixture. (A) Full spectrum of Q1 scan of the PLPC hydroperoxides Ag$^+$ adduct. (B) Silver adducts of PLPC-OOH. (C) Collision-induced dissociation of m/z 896. (D) Fragmentation pathways of 13-HpODE PC.

Ag$^+$ CIS-MS have been applied to identify these novel oxidation products from the oxidation of these phospholipids (Milne, 2002).

 ## 6. Precautions of Applying Ag$^+$ CIS-MS to Study Lipids

Application of Ag$^+$ CIS-MS to identify intact lipid peroxides has been successfully utilized to identify intact lipid peroxides based on their unique fragmentation in the gas phase, but certain precautions are recommended.

First, silver can be deposited in the ion source of a mass spectrometer, and frequent cleaning may be needed. Second, long-time storage of $AgBF_4$ stock solution may cause precipitation, and diluted stock solution is often recommended. $AgNO_4$ can be used to replace $AgBF_4$ when an aqueous mobile phase is used for an LC separation. Last but not the least, the Ag^+ CIS-MS technique is useful for structural identification purposes, and attempts to make it a quantitative method have not been successful primarily because of the formation of multiple silver adducts (Seal *et al.*, 2003). The ratio between the analytes and silver salts turns out to be critical. The numbers of silver binding sites in a molecule—for example, the number of double bonds—is another factor that determines the formation of lipid and silver complexes. For an example, MS response of cholesterol esters is linear when the analyte has the same number of double bonds in the molecule as the internal standard, whereas a nonlinear response is obtained when the analytes and standards have different unsaturation numbers. Complex adducts containing cholesterol esters, silver ion, AgF, $AgBF_4$, and 2-propanoxide are also observed when $AgBF_4$ is in molar excess of the cholesterol esters. These high-molecular-weight adducts lead to a reduced $[M + Ag]^+$ signal. On the other hand, cholesterol esters with more double bonds tend to preferentially bind to silver ion when silver ion is limiting.

7. Conclusions

This chapter reviews the application of silver ion CIS-MS to identify the intact lipid peroxides, including those derived from cholesterol esters and phospholipids. The coordination of silver ion to the lipid peroxides induces characteristic fragmentation in the gas phase of the MS. This fragmentation can be used to identify the structure of peroxides that are otherwise unstable in the MS. A number of novel structures, such as hydroperoxides, monocyclic peroxides, bicyclic endoperoxides, serial cyclic peroxides, and dioxolane-isoprostane peroxides, have been identified from the oxidation of cholesterol arachidonate. Even though this technique is powerful for structure identification purposes, quantification has been difficult.

The preferential formation of intact lipid peroxides can be easily controlled *in vitro*. However, further metabolism of these peroxides *in vivo* has been largely unexplored. Limited research has shown that some of these peroxides can be reduced or rearranged to other biologically active compounds (Musiek *et al.*, 2005). Silver ion CIS-MS may be applied to identify these oxidation products derived from biological sources.

Fish oil has attracted increasing attention because of its beneficiary effects on a number of human diseases. Eicosapentaeoic acid (EPA, C20:5) and docosahexaenoic acid (DHA, C22:6) are two of the major ω-3 fatty acids in fish oil. An increasing amount of evidence indicates that oxidation of these fatty acids may play a role in the biological effects of fish oil. Oxidation of these fatty acids is predicted to be more complex than that of arachidonate because there are more double bonds in the molecule. Thus, more complicated oxidation products are anticipated. Silver-ion CIS-MS can be applied to study the oxidation of EPA and DHA and provide with structure evidence to study the biological significance of these oxidation products (Seal and Porter, 2004; Yin *et al.*, 2005).

ACKNOWLEDGMENTS

The experiments of Christine Havrilla pioneered the application of CIS-MS to the identification of lipid peroxides. We are grateful for her efforts and those of Ginger Milne, who made extensive use of the technique in studies of phospholipid oxidation. We acknowledge stimulating discussions with Jason Morrow in the Division of Clinical Pharmacology, Vanderbilt University School of Medicine. We also thank David Hachey, Wade Calcutt, Lisa Manier, Dawn Overstreet, and Betty Fox of the Mass Spectrometry Research Center of Vanderbilt University. Financial support from National Institutes of Health grants DK 48831, CA 77839, HL17921, GM15431, P30 ES00267, and CHE 9996188 is gratefully acknowledged.

REFERENCES

Bayer, E., Gfrorer, P., and Rentel, C. (1999). Coordination-ionspray-MS (CIS-MS), a universal detection and characterization method for direct coupling with separation techniques. *Angew. Chem. Int. Ed. Engl.* **38**, 992–995.

Blakely, C. R., and Adams, M. J. (1978). Crossed-beam liquid chromatoraph—mass spectrometer combination. *J. Chromatogr.* **158**, 261.

Dole, M., Hines, R. L., Mack, R. C., Mobley, R. C., Ferguson, L. D., and Alice, M. B. (1968). Molecular beams of macrions. *J. Chem. Phys.* **49**, 2240.

Gaskell, S. J. (1997). Electrospray: Principles and practice. *J. Mass Spectrom.* **32**, 677.

Guido, D. M., McKenna, R., and Mathews, W. R. (1993). Quantitation of hydroperoxy-eicosatetraenoic acids and hydroxyeicosatetraenoic acids as indicators of lipid-peroxidation using GC-MS. *Anal. Biochem.* **209**, 123–129.

Hall, L. M., and Murphy, R. C. (1998). Analysis of stable oxidized molecular species of glycerophospholipids following treatment of red blood cell ghosts witht-butylhydroperoxide. *Anal. Biochem.* **258**, 184–194.

Harrison, K. A., Davies, S. S., Marathe, G. K., McIntyre, T., Prescott, S., Reddy, K. M., Falck, J. R., and Murphy, R. C. (2000). Analysis of oxidized glycerophosphocholine lipids using electrospray ionization mass spectrometry and microderivatization techniques. *J. Mass Spectrom.* **35**, 224–236.

Havrilla, C. M., Hachey, D. L., and Porter, N. A. (2000). Coordination (Ag$^+$) ion spray—mass spectrometry of peroxidation products of cholesteryl linoleate and cholesteryl arachidonate: High-performance lipid chromatography—mass spectrometry analysis of peroxide products from polyunsaturated lipid autoxidation. *J. Am. Chem. Soc.* **122**, 8042–8055.

Jiang, Z. Y., Woollard, A. C. S., and Wolff, S. P. (1991). Lipid hydroperoxide measurement by oxidation of Fe^{2+} in the presence of xylenol orange. Comparison with the TBA assay and an iodometric method. *Lipids* **26**, 853–856.

Kikugawa, K. (1997). Use and limitation of thiobarbituric acid (TBA) test for lipid peroxidation. *Recent Res. Dev. in Lipids Res.* **1**, 73–96.

Lee, S. H., Oe, T., and Blair, I. A. (2001). Vitamin C–induced decomposition of lipid hydroperoxides to endogenous genotoxins. *Science* **292**, 2083–2086.

MacMillian, D. K., and Murphy, R. C. (1995). Analysis of lipid hydroperoxides and long-chain conjugated keto acids by negative ion electrospray mass spectrometry. *J. Am. Soc. Mass Spectrom.* **6**, 1190–1201.

Milne, G. L. (2002). "Identification and Analysis of Autoxidation Products of Phospholipids Formed in Human Low Density Lipoproteins." Ph.D. diss. Vanderbilt University.

Milne, G. L., and Porter, N. A. (2001). Separation and identification of phospholipid peroxidation products. *Lipids* **36**, 1265–1275.

Miyazawa, T., Fujimori, K., Suzuki, T., and Yasuda, K. (1994). Determination of phospholipid hydroperoxides using luminol chemilumininescence high-performance liquid-chromatography. *Methods Enzymol.* **233**, 324–332.

Montine, T. J., and Morrow, J. D. (2005). Fatty acid oxidation in the pathogenesis of Alzheimer's disease. *Am. J. Pathol.* **166**, 1283–1289.

Moore, K., and Roberts, L. J., Jr. (1998). Measurement of lipid peroxidation. *Free Radic. Res.* **28**, 659–671.

Morris, L. J. (1966). Separation of lipids by silver ion chromatography. *J. Lipid Res.* **7**, 717–732.

Morrow, J. D. (2005). Quantification of isoprostanes as indices of oxidant stress and the risk of atherosclerosis in humans. *Arterioscl. Thromb. Vas. Biol.* **25**, 279–286.

Morrow, J. D., Hill, E., Burk, R. F., Nammour, T. M., Badr, K. F., and Roberts, L. J., Jr. (1990). A series of prostaglandin-F_2–like compounds are produced *in vivo* in humans by a noncyclooxygenase, free radical-catalyzed mechanism. *Proc. Natl. Acad. Sci. USA* **87**, 9383–9387.

Morrow, J. D., Roberts, I., and Jackson., L. (1999). Mass spectrometric quantification of F_2-isoprostanes in biological fluids and tissues as measure of oxidant stress. *Methods Enzymol.* **300**, 3–12.

Murphy, R. C., Fiedler, J., and Hevko, J. (2001). Analysis of nonvolatile lipids by mass spectrometry. *Chem. Rev.* **101**, 479–526.

Musiek, E. S., Yin, H., Milne, G. L., and Morrow, J. D. (2005). Recent advances in the biochemistry and clinical relevance of the isoprostane pathway. *Lipids* **40**, 987–994.

Nourooz-Zadeh, J., Tajaddini-Sarmadi, J., and Wolff, S. P. (1994). Measurement of plasma hydroperoxide concentrations by the ferrous oxidation-xylenol orange assay in conjunction with triphenylphosphine. *Anal. Biochem.* **220**, 403–409.

Paradis, V., Kollinger, M., Fabre, M., Holstege, A., Poynard, T., and Pondossa, P. (1997). *In situ* detection of lipid peroxidation by-products in chronic liver diseases. *Hepatology* **26**, 135–142.

Poli, G., and Schaur, R. J. (2000). 4-Hydroxynonenal in the pathomechanisms of oxidative stress. *IUBMB Life* **50**, 315–321.

Porter, N. A. (1986). Mechanisms for the autoxidation of polyunsaturated lipids. *Acc. Chem. Res.* **19**, 262–268.

Porter, N. A., Caldwell, S. E., and Mills, K. A. (1995). Mechanisms of free radical oxidation of unsaturated lipids. *Lipids* **30**, 277–290.

Pulfer, M., and Murphy, R. C. (2003). Electrospray mass spectrometry of phospholipids. *Mass Spectrom. Rev.* **22**, 332–364.

Rentel, C., Strohschein, S., Albert, K., and Bayer, E. (1998). Silver-plated vitamins: A method of detecting tocopherols and carotenoids in LC/ESI-MS coupling. *Anal. Chem.* **70**, 4394–4400.

Roberts, L. J., Jr., and Morrow, J. D. (2002). Products of the isoprostane pathway: Unique bioactive compounds and markers of lipid peroxidation. *Cell. Mol. Life Sci.* **59,** 808–820.

Seal, J. R., Havrilla, C. M., Porter, N. A., and Hachey, D. L. (2003). Analysis of unsaturated compounds by Ag$^+$ coordination ionspray mass spectrometry: Studies of the formation of the Ag$^+$/lipid complex. *J. Am. Soc. Mass Spectrom.* **14,** 872–880.

Seal, J. R., and Porter, N. A. (2004). Liquid chromatography coordination ion-spray mass spectrometry (LC-CIS-MS) of docosahexaenoate ester hydroperoxides. *Anal. Bioanal. Chem.* **378,** 1007–1013.

Shen, H. M., and Ong, C. N. (2000). Detection of oxidative DNA damage in human sperm and its association with sperm function and male infertility. *Free Radic. Biol. Med.* **28,** 529–536.

Spickett, C. M., Pitt, A. R., and Brown, A. J. (1998). Direct observation of lipid hydroperoxides in phospholipid vesicles by electrospray mass spectrometry. *Free Radic. Biol. Med.* **25,** 613–620.

Uchida, K. (2000). Role of reactive aldehyde in cardiovascular diseases. *Free Radic. Biol. Med.* **28,** 1685–1696.

Uchida, K., Kanematsu, M., Morimitsu, Y., Osawa, T., Noguchi, N., and Niki, E. (1998). Acrolein is a product of lipid peroxidation reaction. Formation of free acrolein and its conjugate with lysine residues in oxidized low density lipoproteins. *J. Biol. Chem.* **273,** 16058–16066.

Yin, H., Hachey, D. L., and Porter, N. A. (2000). Structural analysis of diacyl peroxides by electrospray tandem mass spectrometry with ammonium acetate: Bond homolysis of peroxide-ammonium and peroxide-proton adducts. *Rapid Commun. Mass Spectrom.* **14,** 1248–1254.

Yin, H., Hachey, D. L., and Porter, N. A. (2001). Analysis of diacyl peroxides by Ag$^+$ coordination ionspray tandem mass spectrometry: Free radical pathways of complex decomposition. *J. Am. Soc. Mass Spectrom.* **12,** 449–455.

Yin, H., Havrilla, C. M., Morrow, J. D., and Porter, N. A. (2002). Formation of isoprostane bicyclic endoperoxides from the autoxidation of cholesteryl arachidonate. *J. Am. Chem. Soc.* **124,** 7745–7754.

Yin, H., Morrow, J. D., and Porter, N. A. (2004). Identification of a novel class of endoperoxides from arachidonate autoxidation. *J. Biol. Chem.* **279,** 3766–3776.

Yin, H., Musiek, E. S., Gao, L., Porter, N. A., and Morrow, J. D. (2005). Regiochemistry of neuroprostanes generated from the peroxidation of docosahexaenoic acid *in vitro* and *in vivo.* *J. Biol. Chem.* **280,** 26600–26611.

Yin, H., and Porter, N. A. (2003). Specificity of the ferrous oxidation of xylenol orange assay: Analysis of autoxidation products of cholesteryl arachidonate. *Anal. Biochem.* **313,** 319–326.

Yin, H., and Porter, N. A. (2005). New insights regarding the autoxidation of polyunsaturated fatty acids. *Antioxid. Redox. Signal.* **7,** 170–184.

Zhang, J., and Brodbelt, J. S. (2005). Silver complexation and tandem mass spectrometry for differentiation of isomeric flavonoid diglycosides. *Anal. Chem.* **77,** 1761–1770.

CHAPTER TWELVE

QUANTIFICATION OF CARDIOLIPIN BY LIQUID CHROMATOGRAPHY-ELECTROSPRAY IONIZATION MASS SPECTROMETRY

Teresa A. Garrett, Reza Kordestani, *and* Christian R. H. Raetz

Contents

Abstract

Cardiolipin (CL), a tetra-acylated glycerophospholipid composed of two phosphatidyl moieties linked by a bridging glycerol, plays an important role in mitochondrial function in eukaryotic cells. Alterations to the content and acylation state of CL cause mitochondrial dysfunction and may be associated with pathologies such as ischemia, hypothyrodism, aging, and heart failure. The structure of CL is very complex because of microheterogeneity among its four acyl chains. Here we have developed a method for the quantification of CL molecular species by liquid chromatography-electrospray ionization mass spectrometry. We quantify the $[M-2H]^{2-}$ ion of a CL of a given molecular formula and identify the CLs by their total number of carbons and unsaturations in the acyl chains. This method, developed using mouse macrophage RAW 264.7 tumor cells, is broadly applicable to other cell lines, tissues, bacteria and yeast. Furthermore, this method could be used for the quantification of lyso-CLs and bis-lyso-CLs.

Department of Biochemistry, Duke University Medical Center, Durham, North Carolina

Methods in Enzymology, Volume 433
ISSN 0076-6879, DOI: 10.1016/S0076-6879(07)33012-7

1. INTRODUCTION

Cardiolipin (CL) is a tetra-acylated glycerophospholipid composed of two phosphatidyl moieties linked by a bridging glycerol (Fig. 12.1). This unique lipid possesses up to three chiral centers (Fig. 12.1A) and can be acylated with a variety of different acyl chains. Because of this structure, the number of possible CL molecular species is staggering. With 10 different acyl chains randomly distributed among the four esters it has been estimated

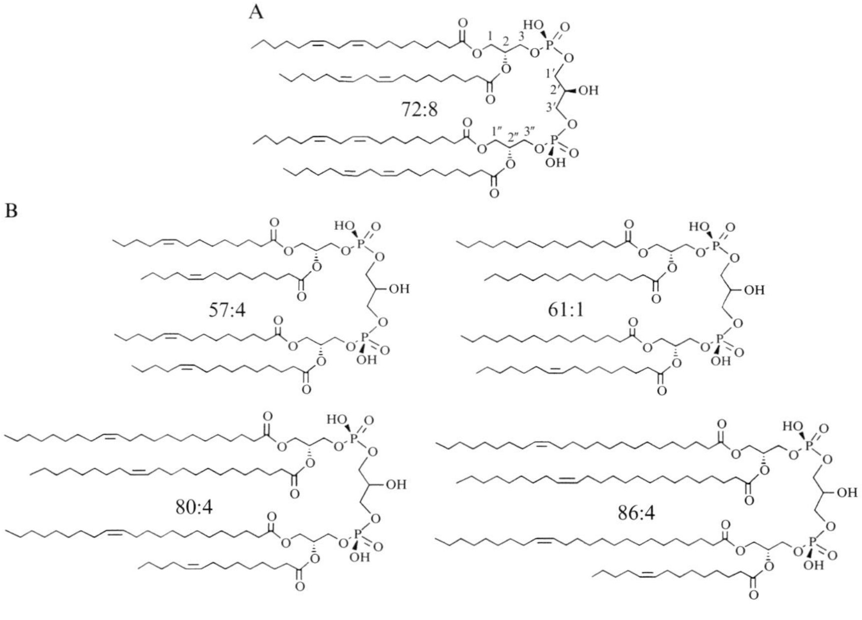

Figure 12.1 Structure of CL. (A) The structure of tetra-linoleoyl CL is shown as a representative CL structure. The glycerol carbons are numbered as shown to allow clear reference to each of the carbons. The three possible chiral centers are the 2, 2′, and 2″ carbons. If the phosphatidyl groups are the same, as shown here, then the central carbon of the bridging glycerol (2′) is not chiral. (B) The structures of the four synthetic CL standards are shown. As indicated in the structures, the 2′ carbon is racemic in the synthetic standards. The designation of the standards given below uses nomenclature adopted by the LipidMAPS consortium (www.lipidmaps.org) (Fahy *et al.*, 2005). The 57:4 standard is 1′-[1,2-di-(9Z-tetradecenoyl)-*sn*-glycero-3-phospho, 3′-[1-(9Z-tetradecenoyl), 2-(10Z-pentadecenoyl)-*sn*-glycero-3-phospho]-*sn*-glycerol. The 61:1 standard is 1′-[1,2-dipentadecanoyl-*sn*-glycero-3-phospho], 3′-[1-(pentadecanoyl), 2-(9Z-hexadecenoyl)-*sn*-glycero-3-phospho]-*sn*-glycerol. The 80:4 standard is 1′-[1,2-di-(13Z-docosenoyl)-*sn*-glycero-3-phospho], 3′-[1-(13Z-docosenoyl), 2-(9Z-tetradecenoyl)-*sn*-glycero-3-phospho]-*sn*-glycerol. The 86:4 standard is 1′-[1,2-di-(15Z-tetracosenoyl)-*sn*-glycero-3-phospho], 3′-[1-(15Z-tetracosenoyl), 2-(9Z-tetradecenoyl)-*sn*-glycero-3-phospho]-*sn*-glycerol.

that 10^4 distinct CL molecular structures can be envisioned (Schlame et al., 2000). As discussed below, the cell typically does not possess all of these possible combinations. The acyl chain composition of CL is regulated and remodeled depending on the organism, organ, and cell type (Han et al., 2006; Mutter et al., 2000; Schlame et al., 2005; Sparagna et al., 2005; Xu et al., 2006).

Cardiolipin is found in both prokaryotic and eukaryotic cells. It is localized in membranes involved in the generation of an electrochemical gradient used for the production of ATP and the transport of substrates across the membrane (Schlame et al., 2000). In eukaryotic cells, CL is highly enriched in mitochondria, especially the inner mitochondrial membrane (Krebs et al., 1979). It binds to cytochrome c oxidase tightly and is necessary for efficient transport of electrons through this protein complex (Robinson, 1993). In addition, it is associated with the ATP/ADP exchange protein (Beyer and Klingenberg, 1985), the F_0F_1 ATP synthase (Eble et al., 1990), and the cytochrome bc_1 complex (Hayer-Hartl et al., 1992; Schlame et al., 2000). CL may play a role in apoptosis through involvement in cytochrome c release from the mitochondria (Gonzalvez and Gottlieb, 2007).

Cardiolipin is synthesized utilizing a similar pathway as other glycerophospholipids (Heath et al., 2002; Vance, 2002) (Fig. 12.2). Glycerol 3-phosphate is acylated to phosphatidic acid and then activated to CDP-diacylglycerol (CDP-DAG). A second glycerol 3-phosphate reacts with CDP-DAG to form phosphatidylglycerol phosphate. A phosphatase removes the phosphate from the terminal glycerol to form phosphatidylglycerol. Up to this point the eukaryotic and prokaryotic biosynthetic pathways are analogous. The final step leading to the formation of CL differs between prokaryotes and eukaryotes. In eukaryotes, phosphatidylglycerol reacts with a second CDP-diacylglycerol molecule to form CL and release CMP (see Fig. 12.2) (Schlame and Greenberg, 1997; Schlame and Hostetler, 1997; Schlame et al., 1993; Tamai and Greenberg, 1990). In prokaryotes, two phosphatidylglycerol molecules react to form CL and release free glycerol (see Fig. 12.2) (Schlame et al., 2000). Details of the genetics and regulation of CL biosynthesis have been reviewed (Schlame et al., 2000).

In mammalian cells, the fatty acyl chains of CL are enriched in 18 carbon unsaturated fatty acids (Hoch, 1992; Schlame et al., 2005). Lineolic acid (18:2) is particularly abundant in CL of mammalian cells, comprising 80 to 90% of the acyl chains in CL isolated from mammalian heart (Schlame et al., 2005). This specific composition is obtained primarily via remodeling of the CL acyl chains following de novo synthesis. CL remodeling occurs via two distinct pathways, following deacylation to mono-lysoCL (MLCL) (Hatch, 1998). A mitochondrial phospholipase A2 may play a role in MLCL formation (Buckland et al., 1998). Next, MLCL can be re-acylated by acyl-CoA–dependent and acyl-CoA–independent pathways. An acyl-CoA–dependent MLCL acyltransferase activity (Ma et al., 1999)

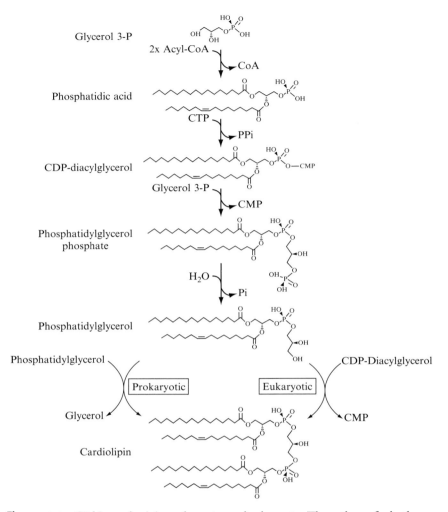

Figure 12.2 CL biosynthesis in prokaryotes and eukaryotes. The pathway for both pro-karyotes and eukaryotes is similar until the final step (Heath *et al.*, 2002; Vance, 2002). The prokaryotic CL synthase combines two phosphatidylglycerol molecules to form CL. In contrast, the eukaryotic CL synthase reacts phosphatidylglycerol with CDP-dia-cylglycerol (CDP-DAG) to form CL. More details of the genetics, regulation, and metabolism of CL are found in a recent review by Schlame *et al.* (2000).

found in the mitochondria acylates MLCL preferentially with unsaturated acyl chains (Taylor and Hatch, 2003). However, this acyl-CoA–dependent MLCL acyltransferase does not have the clear specificity for 18:2 acyl-CoA required for the enrichment of linoleolate in mammalian CL (Schlame *et al.*, 2005). A second re-acylation pathway involves a transacylation reaction in which acyl chains are transferred from a glycerophospholipid, such as phopsphatidylcholine or phosphatidylethanolamine, to MLCL. This enzyme is selective for 18:2 acyl chains and has recently been identified as tafazzin (Xu *et al.*, 2006).

Alterations in CL content and acylation state are associated with a variety of pathological conditions including ischemia, hypothyrodism, aging, and heart failure (Chicco and Sparagna, 2007; Hauff and Hatch, 2006). In these pathological states, CL alterations are correlated to mitochondrial dysfunction (Chicco and Sparagna, 2007). In addition, a cardiomyopathy, known as Barth syndrome—characterized by skeletal muscle weakness, neutropenia, and growth retardation (Gu *et al.*, 2004)—is attributed to defects in CL metabolism. The genetic defect that causes Barth syndrome was mapped to the X chromosome (Ades *et al.*, 1993) and identified as the tafazzin gene (Barth *et al.*, 1999). As discussed above, tafazzin is the transacylase involved in the remodeling of CL (Xu *et al.*, 2006). Biopsies of muscle from Barth syndrome patients reveal mitochondria with abnormal structures and respiratory function (Barth *et al.*, 1983). Analysis of CL from affected tissues of Barth syndrome patients shows a deficiency in total CL levels with a specific deficiency in tetra-linoleoyl CL (see Fig. 12.1) and an increase in the levels of MLCL (Hauff and Hatch, 2006; Valianpour *et al.*, 2005). Tetra-linoleoyl CL is the most abundant CL species in heart and skeletal muscle (Schlame and Ren, 2006) and, therefore, it is the loss of this molecular species that may contribute to the clinical manifestations of Barth syndrome. The loss of tetra-linoleoyl CL is consistent with the loss of this linoleoyl-selective transacylase activity. Interestingly, tetra-linoleoyl CL deficiency is considered a specific marker of Barth syndrome and is used in the differential diagnosis of the disease (Schlame and Ren, 2006; Valianpour *et al.*, 2002).

Herein we present new methods for the detection and quantification of CL using liquid chromatography-electrospray ionization mass spectrometry (LC-ESI-MS). While the methods reported here were developed using mouse macrophage RAW 264.7 tumor cells (Raetz *et al.*, 2006), we have found that they are broadly applicable to other cell lines and, tissues. While here we present the quantification of tetra-acylated CL, these methods are, in principle, applicable to the quantification of lyso-CL and bis-lyso-CL.

2. Complexity of Cardiolipin Structure

A given molecular mass of CL can represent several distinct molecular structures. For example, given a CL with three linoleoyl (18:2) and one oleoyl (18:1) chain, four different molecular structures are possible. In one structure, the 18:1 chain could occupy the 1 position with the 18:2 chains on the 2, 1″, and 2″ positions. In a second structure, the 18:1 chain could occupy the 2 position with the 18:2 chains on the 1, 1″, and 2″ positions. The remaining two possible structures would have the 18:1 in the 1′ and 2″ positions, respectively, with the 18:2 chains in the other positions. MS analysis cannot distinguish among these different molecular structures. They have the exact same molecular formula ($C_{81}H_{144}O_{17}P_2$) and exact mass ([M] = 1450.988). In addition, this molecular formula and exact mass are shared by CL molecules with different acyl chain composition. A CL with two 18:2 chains, one arachidonyl (20:4) chain, and one palmitoyl acyl chain (16:0) would have the same molecular formula and exact mass as the (18:2)$_3$, 18:1 CL. Therefore, a given molecular mass of CL can represent a number of different molecular structures. Because of this structural complexity, we represent the CL by the total number of carbons and unsaturations in the acyl chains. The molecular formula, $C_{81}H_{144}O_{17}P_2$, would be called 72:7 CL. Our method cannot distinguish the various molecular species that contribute to a CL ion of a given mass.

3. ANALYSIS OF SYNTHETIC CARDIOLIPIN STANDARDS USING LIQUID CHROMATOGRAPHY-MASS SPECTROMETRY

When analyzed by negative ion electrospray ionization mass spectrometry, CL is detected predominantly as a doubly charged ion [M-2H]$^{2-}$, although the [M-H]$^{1-}$ ion can be observed at a lower abundance. These doubly charged ions are in the same mass range (m/z 670–780) as many of the abundant singly charged ions of glycerophospholipids such as phosphatidic acid, phosphatidylethanolamine, and phosphatidylglycerol. Pre-fractionation using anion exchange, reverse-phase, or normal-phase chromatography is required to separate the CL from these abundant, efficiently ionizing lipid species. While we have utilized all three chromatography methods to analyze CLs, in this chapter we will focus on an LC-MS method that utilizes a Zorbax SB-C8 reverse-phase column (5 μm, 2.1 × 50 mm, Agilent, Palo Alto, CA) coupled to a QSTAR XL quadropole, time-of-flight tandem mass spectrometer (Applied Biosystems/MDS Sciex). The specific details of the chromatography and negative-ion MS are given in our accompanying chapter in the previous volume (Garrett et al., 2007).

In order to quantify the CL levels in the macrophage RAW 264.7, we first needed quantitative synthetic standards. Four CL standards were synthesized by Avanti Polar Lipids. The structures of the standards are shown in Fig. 12.1B. Table 12.1 shows the predicted and observed masses of the [M-H]$^{1-}$ and [M-2H]$^{2-}$ of the standards. These standards were chosen because they are not isobaric with other CL molecules or other negatively charged molecules that elute with similar retention times. They are both smaller and larger than the masses of the CLs found in the RAW cell. Other cells and tissues may have CLs with slightly different chain lengths making some of the standards inappropriate for quantification. For all of the tissues and cells that we have analyzed, at least three of the four standards have been useful for quantification.

To characterize the behavior of the standards using LC-MS each of the standards was diluted into mobile phase A (Garrett et al., 2007) at a concentration of 40, 80, 200, 500, 1000, and 2000 nM. Next, 25 μl of each dilution was injected, in turn, onto a Zorbax SB-C8 reverse-phase column at 200 μl/min and eluted and analyzed as described in Garrett et al. (2007).

Figure 12.3A shows the extracted ion current (EIC) for the [M-2H]$^{2-}$ ions of the standards. For this analysis, each was analyzed on separate LC-MS runs and the spectra overlaid to show the difference in retention

Table 12.1   Masses of cardiolipin standards

Cardiolipin Standard	Formula [M]	[M-H]$^{1-}$ Observed (*m/z*)	Exact (*m/z*)	[M-2H]$^{2-}$ Observed (*m/z*)	Exact (*m/z*)
57:4	$C_{66}H_{120}O_{17}P_2$	1245.789	1245.792	622.390	622.392
61:1	$C_{70}H_{134}O_{17}P_2$	1307.890	1307.902	653.444	653.447
80:4	$C_{89}H_{166}O_{17}P_2$	1568.148	1568.152	783.570	783.573
86:4	$C_{95}H_{178}O_{17}P_2$	1652.238	1652.247	825.617	825.620

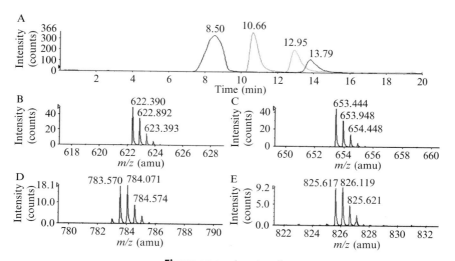

Figure 12.3   (*continued*)

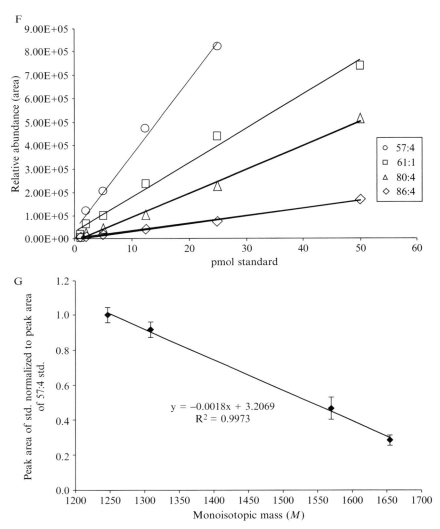

Figure 12.3 Negative-ion LC-MS analysis of synthetic CL standards. (A) One picomole of each synthetic CL standard was analyzed by reverse-phase LC-MS (Garrett *et al.*, 2007). The extracted ion current (EIC) for the $[M-2H]^{2-}$ ions of the four synthetic standards are shown overlaid. The 57:4 standard elutes at min 8.5, followed by the 61:1 standard at 10.66 min, the 80:4 standard at 12.95 min and the 86:4 standard at 13.79 min. (B) The mass spectrum of the material eluting from min 7.9 to 9.1, representing the $[M-2H]^{2-}$ ion of the 57:4 standard. (C) The mass spectrum of the material eluting from min 10.3 to 11.2, representing the $[M-2H]^{2-}$ ion of the 61:1 standard. (D) The mass spectrum of the material eluting from min 12.7 to 13.6, representing the $[M-2H]^{2-}$ ion of the 80:4 standard. (E) The mass spectrum of the material eluting from min 13.4 to 14.2, representing the $[M-2H]^{2-}$ ion of the 86:4 standard. (F) The plot of the peak area of the EIC of the $[M-2H]^{2-}$ ion of each of the standards versus the pmol of standard injected onto the Zorbax C-8 reverse phase column (Garrett *et al.*, 2007). The 57:4 CL standard was linear from 1 to 25 pmol (equation for line).

times. As expected, the 57:4 standard eluted first with a retention time of approximately 8.5 min, followed by the 61:1 standard at 10.66 min, the 80:4 standard at 12.95 min, and finally, the 86:4 standard at 13.79 min. The mass spectra for the $[M-2H]^{2-}$ ions of the standards are shown in Fig. 12.3B through E. The collision-induced dissociation spectra of the $[M-2H]^{2-}$ ions revealed fragment ions consistent with the proposed structures and published fragmentation patterns (data not shown) (Hoischen *et al.*, 1997; Pulfer and Murhpy, 2003). The peak areas of the extracted ion current (EIC) for the 61:1, 80:4, and 86:4 standards were linear for 1 pmol to 50 pmol injected onto the column, as shown in Fig. 12.3F. The peak area of the EIC for the 57:4 standard was linear for 1 pmol to 25 pmol injected onto the column. In general, approximately 5 to 10 pmol of standard is injected during sample analysis using this method. The mass spectrometry response of the four standards decreased with increasing chain length. Figure 12.3G shows the plot of the peak area of the EIC for each of the standards normalized to the peak area of the EIC of the 57:4 standard versus molecular weight. This plot shows that the mass spectrometry response decreases linearly with increasing chain length. From this plot we can generate an ionization correction factor to quantify CL molecules with a broad range of molecular masses.

4. Liquid Chromatography-Mass Spectrometry of Cardiolipins in Total Lipid Extracts

Total lipid extracts from RAW cells were prepared as described previously (Garrett *et al.*, 2007). Using the LC-MS system described above, we were able to quantify nine species of CL in these extracts (Table 12.2). The CLs are identified by the number of carbons and unsaturations in the acyl chain component. The exact molecular structure of the CL (i.e., the exact location and identity of the four acyl chains) cannot be determined using this method. However, LC-MS/MS analysis can give a qualitative indication of the predominant fatty acyl chain composition within the isobaric mixture.

We determined the amount of CL standards to add to the extraction mixture in order to have the peak area of the EIC for the standards be about equal to the peak area of the EIC for the CLs being quantified in the sample.

The 61:1, 80:4, and 86:4 CL standards were linear from 1 to 50 pmol. (G) The peak areas of each of the standards were divided by the peak area for the 57:4 CL standard. The resulting ratio is plotted versus the molecular weight of the standards. The error bars are the standard error of the mean from 10 separate LC-MS runs.

Table 12.2   Cardiolipins quantified from RAW cells

Cardiolipins	Formula [M]	$[M-2H]^{2-}$ Observed (m/z)	Exact (m/z)
68:5	$C_{77}H_{140}O_{17}P_2$	698.510	698.473
68:4	$C_{77}H_{142}O_{17}P_2$	699.504	699.480
68:2	$C_{77}H_{146}O_{17}P_2$	701.516	701.495
70:6	$C_{79}H_{142}O_{17}P_2$	711.503	711.479
70:5	$C_{79}H_{144}O_{17}P_2$	712.508	712.486
70:4	$C_{79}H_{146}O_{17}P_2$	713.513	713.493
72:6	$C_{81}H_{146}O_{17}P_2$	725.516	725.496
72:5	$C_{81}H_{144}O_{17}P_2$	726.524	726.603
72:4	$C_{81}H_{142}O_{17}P_2$	727.529	727.510

By varying the amount of CL standards added during the co-extraction, we determined that for a 150-mm culture dish grown to ∼90% confluence, 500 pmol of each of the CL standards gave a peak area of the EIC about equal to the peak area of the EIC for the major CL species of the cell. The standards were synthesized by Avanti Polar lipids and delivered as a mixture of the four standards, each at 10 μM.

To summarize the method thus far, 500 pmol of each of the CL standard is added to the cell suspension (1 ml) before the addition of chloroform (1.25 ml) and methanol (2.5 ml) to form a single-phase extraction mixture. Next, the lipids are recovered by conversion to a two-phase system, as described previously (Garrett *et al.*, 2007), dried under a stream of nitrogen and stored at $-20°$ until analysis. The dried lipids are re-dissolved in 0.2 ml of chloroform:methanol (1:1, v:v) and then diluted five fold into mobile phase A (Garrett *et al.*, 2007). Ten microliters are injected onto the reverse-phase column and analyzed using the LC-MS procedure described previously (Garrett *et al.*, 2007).

As stated above, the mass range that the $[M-2H]^{2-}$ CL ions occupy overlaps with the mass range of other abundant phospholipids. For example, in the reverse-phase system used here, CL elutes with phosphatidylethanol-amine species. As shown in Fig. 12.4, doubly charged 72:5 CL (predicted m/z 726.603) is isobaric with a singly charged species predicted to be a 1 Z-alkenyl, 2-acylglycerophosphoethanolamine with 36 carbons, and 2 unsa-turations in the acyl chains (36:2 plasmenyl-ethanolamine [PLE], predicted m/z 726.544) (Fahy *et al.*, 2005) by exact mass and retention time on the column. When normal-phase chromatography (Becart *et al.*, 1990) is used instead of reverse-phase chromatography, the doubly charged 76:12 CL ion

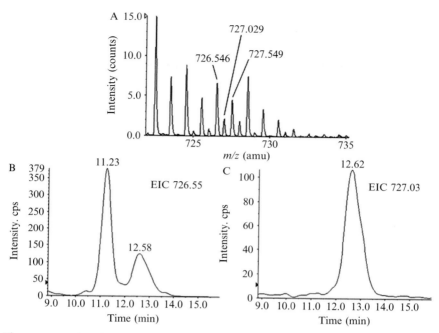

Figure 12.4 Overlap of doubly charged CL ions by singly charged ions derived from other phospholipids. (A) The mass spectrum of the material eluting from min 10.8 to 13.9. The doubly charged 72:5 CL is identified by the $^{13}C_1$ isotope peak at m/z 727.029. A singly charged ion appears to overlap the predicted monoisotopic peak at m/z 726.546. (B) The extracted ion current for the monoisotopic 72:5 CL, m/z 726.55. (C) The extracted ion current for the $^{13}C_1$ isotope for 72:5 CL, m/z 727.03. When the $^{13}C_1$ isotope is extracted from the total ion current the peak has an area that is solely attributable to the 72:5 CL.

(predicted m/z 747.479) elutes with a singly charged species predicted to be the 34:1 phosphatidylglycerol ion (predicted m/z 747.518) (data not shown). The CL ions are distinguished from the overlapping singly charged ions by the half mass unit difference between the monoisotopic $[M-2H]^{2-}$ ion and the ion with one ^{13}C atom (abbreviated as $^{13}C_1$). In Fig. 12.4A, the monoisotopic $[M-2H]^{2-}$ ion of 72:5 CL is nearly isobaric with the $[M-H]^{1-}$ ion of 36:2 PLE. However, the CL-$^{13}C_1$ ion, m/z 727.029, is not overlapped by the either the monoisotopic $[M-H]^{1-}$ PLE ion or the PLE-$^{13}C_1$ ion. By analyzing the $[M-2H]^{2-}$ CL containing one ^{13}C atom we can identify readily CLs among other singly charged ions. In addition, we can determine a CL-specific EIC during LC-MS that can be used to determine peak area. When the monoisotopic mass $[M-2H]^{2-}$ for 72:5 CL is extracted from the total ion current chromatograph, the result is shown in Fig. 12.4B. The peak at 11.23 min corresponds to singly charged ions, not CL. The smaller side peak at 12.58 min corresponds to 72:5 CL. If the mass of the $^{13}C_1$ ion for 72:5 CL is extracted, then a peak at 12.62 min is isolated and can be used for quantification. Using this method we can analyze CL among other singly charged ions without the development of elaborate chromatography methods.

An additional complication for the quantification of CL is the significant overlap of the monoisotopic $[M-2H]^{2-}$ ions and the ^{13}C isotope ions among the various species of CL. Figure 12.5 illustrates how the predicted ^{13}C isotopes overlap among 72:8, 72:7, and 72:6 CLs. The $^{13}C_2$ isotope ion of 72:8 CL (marked with * in Fig. 12.5A), overlaps with the monoisotopic ion of the 72:7 CL (marked with * in Fig. 12.5B). Furthermore, the $^{13}C_2$ isotope ion of the 72:7 CL (marked with ◆ in Fig. 12.5B) overlaps with the monoisotopic ion of CL 72:6 (marked with ◆ in Fig. 12.5C). When these spectra are summed (Fig. 12.5D), the prediction is very similar to the actual mass spectrum (Fig. 12.5E). Because of the large number of carbons in CL the $^{13}C_2$ isotope ion can be more than 50% of the peak area of the most abundant isotope and, therefore, cannot be disregarded.

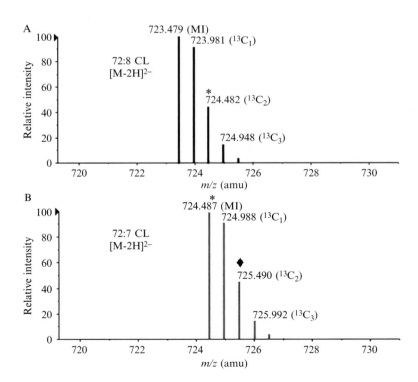

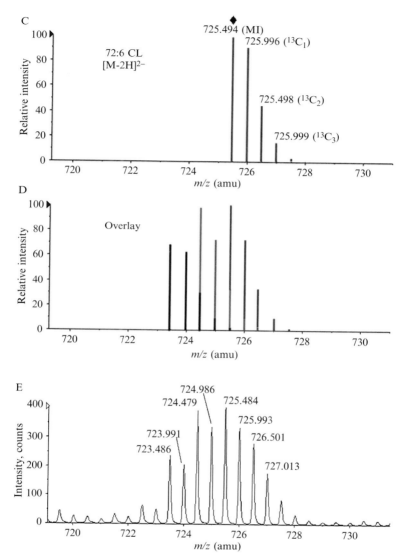

Figure 12.5 Overlap of ^{13}C isotopes. The predicted isotopic distribution of 72:8, 72:7 and 72:6 CL are shown to illustrate the overlap of ^{13}C isotopes and how they complicate the interpretation of the mass spectrum. (A) through (C) show the predicted isotopic distribution for the [M-2H]$^{2-}$ ion of the 72:8, 72:7, and 72:6 CLs. The monoisotopic ion is labeled MI. The remaining peaks are the major ^{13}C isotopes. The isotope with one ^{13}C is marked ^{13}C$_1$, the isotope with two ^{13}C's is marked ^{13}C$_2$, and so on. (D) Result of summing the spectra from (A) and (B). (E) Corresponding region of the mass spectrum of the material eluting from min 12.2 to 15.4 of a LC-MS analysis of total lipid extracts from mouse heart. A normal-phase chromatography system was used (Becart *et al.*, 1990).

5. Quantification of Cardiolipin

We chose to use the $^{13}C_1$ isotope ion to quantify CL. This allows us to distinguish the doubly charged CLs from other singly charged ions of the similar monoisotopic masses as discussed above (see Fig. 12.4). For each CL to be quantified, we needed to determine what the peak area of the EIC for the monoisotopic CL ion would be if there was no overlap of other ions or other CL ^{13}C isotopes. For the case where there is no overlap of CL ^{13}C isotopes, such as with the 72:8 CL (see Fig. 12.5), we first determined the peak area of the EIC for the $^{13}C_1$ ion (Fig. 12.6). Next, the peak area of the monoisotopic ion was back–calculated using the predicted ratio of the different ^{13}C isotopes (see Fig. 12.6). Table 12.3 shows the predicted isotopic distributions for the 72:8 CL calculated using the isotopic distribution calculator of the Analyst QS software (ABI/MDS Sciex). The $^{13}C_1$ isotope ion is 92% of the peak area of the EIC of the monoisotopic ion. This calculated monoisotopic peak area is used for the final quantification below.

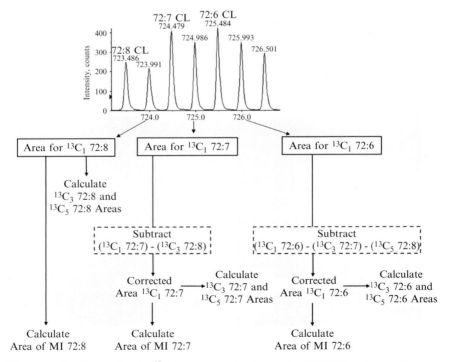

Figure 12.6 Correcting for ^{13}C isotope overlap. Using the 72:8, 72:7 and 72:6 CLs as an example, the method for deconvoluting the ^{13}C isotope overlap is shown. "Area" represents the peak area of the EIC. MI, monoisotopic.

Table 12.3 Isotope distribution for 72:8 cardiolipin

Isotope	72:8 Cardiolipin
MI	1
$^{13}C_1$	0.922
$^{13}C_2$	0.454
$^{13}C_3$	0.157
$^{13}C_4$	0.043
$^{13}C_5$	0.009

MI, monoisotopic.

As shown in Fig. 12.5, the contribution of ^{13}C isotopes to the peak area of adjacent CLs cannot be ignored in quantification calculations. We will illustrate our method for dealing with this problem using the overlapping 72:8, 72:7, and 72:6 CLs (see Figs. 12.5 and 12.6). The peak area of the EIC for the $^{13}C_1$ isotope of the 72:8 CL is determined as above. The peak areas of the $^{13}C_3$ and $^{13}C_5$ isotopes for 72:8 CL are calculated using the isotopic distributions (see Table 12.3, Fig. 12.6). Next, the peak area of the EIC for the $^{13}C_1$ isotope for 72:7 CL is determined. The calculated peak area of the $^{13}C_3$ isotope for 72:8 CL is subtracted from the calculated peak area of the $^{13}C_1$ isotope of the 72:7 CL to generate a corrected peak area for the $^{13}C_1$ isotope for 72:7 CL. This corrected peak area is used to calculate the peak area of the EIC for the monoisotopic 72:7 CL and the peak area of the EIC for the $^{13}C_3$ and $^{13}C_5$ isotopes. For the 72:6 CL, the peak area of the EIC for the $^{13}C_1$ isotope is determined. The peak area of the $^{13}C_3$ isotope for the 72:7 CL and the $^{13}C_5$ isotope for the 72:8 CL are then subtracted from the peak area of the $^{13}C_1$ isotope for 72:6 CL to generate a corrected peak area for the $^{13}C_1$ isotope for 72:6 CL. The process can be continued by calculating the peak area of the EIC for the $^{13}C_3$ and $^{13}C_5$ isotopes for 72:6. These isotopes can be used for the correction of less unsaturated CLs, such as 72:5 and 72:4 CLs. Finally, the corrected peak areas of the $^{13}C_1$ isotopes for the 72:8, 72:7, and 72:6 CLs are used to calculate the peak area of the monoisotopic ion, again, using the isotope distribution specific to that molecular formula.

To convert the corrected monoisotopic peak areas generated above to picomoles of CL per sample, we utilize the standards added during extraction. The peak area of the EIC for each standard is determined using the $^{13}C_1$ isotope as described above for the 72:8 CL. The isotopic distribution is used to calculate the peak area of the EIC for the monoisotopic ion. Generally, there is no overlap of the standards with other ions. We utilize the $^{13}C_1$ isotope for the quantification and for consistency. Because the MS response of the CL standards decreases with increasing molecular weight, we generated a normalization curve (see Fig. 12.3F and G). The peak area of each standard is normalized to the peak area of one of the standards. We have normalized to the 57:4 standard as shown in Fig. 12.3G. The slope of the line reflects the relationship between molecular weight and ionization efficiency relative to the 57:4 standard. The ionization efficiency, for a given molecular weight CL, is multiplied by the peak area of the EIC for the 57:4 standard. For example, the peak area of the EIC for the 57:4 standard would be multiplied by 0.59 for the analysis of 72:8 CL. This effectively decreases the peak area of the EIC for the 57:4 standard to match the decrease in MS response with increasing carbons in the acyl chains.

The final calculation for quantifying 72:8 CL, for example, is as follows: [72:8 CL peak area / (57:4 standard peak area × 0.59)] × 500 pmol = pmol 72:8 in sample.

This method for quantifying CL utilizes the ability to resolve the ^{13}C isotopes of the doubly charged CL molecules using a high-resolution, quadropole time-of-flight MS. Further chromatography could alleviate the overlap of the CL with other singly charged ions such as phosphatidyl-ethanolamine and phosphatidylglycerol. However, the CLs with different degrees of unsaturation cannot be separated using standard LC systems. Our method addresses both of these issues to allow the quantification of a large range of CL molecular species.

ACKNOWLEDGMENTS

We thank Dr. Robert Murphy for review of the manuscript. This work was supported by the LIPID MAPS Large Scale Collaborative Grant GM-069338 from the National Institutes of Health.

REFERENCES

Ades, L. C., Gedeon, A. K., Wilson, M. J., Latham, M., Partington, M. W., Mulley, J. C., Nelson, J., Lui, K., and Sillence, D. O. (1993). Barth syndrome: Clinical features and confirmation of gene localisation to distal Xq28. *Am. J. Med. Genet.* **45,** 327–334.

Barth, P. G., Scholte, H. R., Berden, J. A., van der Klei-van der Harten, J. J., and Sobotka-Plojhar, M. A. (1983). An X-linked mitochondrial disease affecting cardiac muscle, skeletal muscle and neutrophil leucocytes. *J. Neurol. Sci.* **62,** 327–355.

Barth, P. G., Wanders, R. J., Vreken, P., Janssen, E. A., Lam, J., and Baas, F. (1999). X-linked cardioskeletal myopathy and neutropenia (Barth syndrome) (MIM 302060). *J. Inherited Metab. Disord.* **22,** 555–567.

Becart, J., Chavalier, C., and Biesse, J. P. (1990). Quantitative analysis of phospholipids by HPLC with light scattering detector: Application to raw materials for cosmetic use. *J. High Res. Chromatogr.* **13,** 126–129.

Beyer, K., and Klingenberg, M. (1985). ADP/ATP carrier protein from beef heart mito-chondria has high amounts of tightly bound cardiolipin, as revealed by ^{31}P nuclear magnetic resonance. *Biochemistry* **24,** 3821–3826.

Buckland, A. G., Kinkaid, A. R., and Wilton, D. C. (1998). Cardiolipin hydrolysis by human phospholipase A2. The multiple enzymatic activites of human cytosolic phospholipase A2. *Biochim. Biophys. Acta* **1390,** 65–72.

Chicco, A. J., and Sparagna, G. C. (2007). Role of cardiolipin alterations in mitochondrial dysfunction and disease. *Am. J. Cell Physiol.* **292,** C33–C44.

Eble, K. S., Coleman, W. B., Hantgan, R. R., and Cunningham, C. C. (1990). Tightly associated cardiolipin in the bovine heart mitochondrial ATP synthase as analyzed by ^{31}P nuclear magnetic resonance spectroscopy. *J. Biol. Chem.* **265,** 19434–19440.

Fahy, E., Subramaniam, S., Brown, H. A., Glass, C. K., Merrill, A. H. J., Murphy, R. C., Raetz, C. R., Russell, D. W., Seyama, Y., Shaw, W., Shimizu, T., Spener, F., *et al.* (2005). A comprehensive classification system for lipids. *J. Lipid Res.* **46,** 839–861.

Gonzalvez, F., and Gottlieb, E. (2007). Cardiolipin: Setting the beat of apoptosis. *Apoptosis* **12,** 877–885.

Gu, Z., Valianpour, F., Chen, S., Vaz, F. M., Hakkaart, G. A., Wanders, R. J., and Greenberg, M. L. (2004). Aberrant cardiolipin metabolism in the yeast taz1 mutant: A model for Barth syndrome. *Mol. Microbiol.* **51,** 149–158.

Han, X., Yang, K., Yang, J., Cheng, H., and Gross, R. W. (2006). Shotgun lipidomics of cardiolipin molecular species in lipid extracts of biological samples. *J. Lipid Res.* **47,** 864–879.

Hatch, G. M. (1998). Cardiolipin: Biosynthesis, remodeling and trafficking in the heart and mammalian cells. *Int. J. Mol. Med.* **1,** 33–41.

Hauff, K. D., and Hatch, G. M. (2006). Cardiolipin metabolism and Barth syndrome. *Prog. Lipid Res.* **45,** 91–101.

Hayer-Hartl, M., Schagger, H., von Jagow, G., and Beyer, K. (1992). Interactions of phospholipids with the mitochondrial cytochrome-c reductase studied by spin-label ESR and NMR spectroscopy. *Eur. J. Biochem.* **209,** 423–430.

Heath, R. J., Jackowski, S., and Rock, C. O. (2002). Fatty acid and phospholipid metabolism in prokaryotes. *In* "Biochemistry of Lipids, Lipoproteins and Membranes" (D. E. Vance and J. E. Vance, eds.), Vol. 36. Elsevier, New York. pp. 55–92.

Hoch, F. L. (1992). Cardiolipins and biomembrane function. *Biochim. Biophys. Acta* **1113,** 71–133.

Hoischen, C., Ihn, W., Gura, K., and Gumpert, J. (1997). Structural characterization of molecular phospholipid species in cytoplasmic membranes of the cell wall-less *Streptomyces hygroscopicus* L form by use of electrospray ionization coupled with collision-induced dissociation mass spectrometry. *J. Bacteriol.* **179,** 3437–3442.

Krebs, J. J., Hauser, H., and Carafoli, E. (1979). Asymmetric distribution of phospholipids in the inner membrane of beef heart mitochondria. *J. Biol. Chem.* **254,** 5308–5316.

Ma, B. J., Taylor, W. A., Dolinsky, V. W., and Hatch, G. M. (1999). Acylation of monolysocardiolipin in rat heart. *J. Lipid Res.* **40,** 1837–1845.

Mutter, T., Dolinsky, V. W., Ma, B. J., Taylor, W. A., and Hatch, G. M. (2000). Thyroxine regulation of monolysocardiolipin acyltransferase activity in rat heart. *Biochem. J.* **346,** 403–406.

Pulfer, M., and Murhpy, R. C. (2003). Electrospray mass spectrometry of phospholipids. *Mass Spectrom. Rev.* **22,** 332–364.

Raetz, C. R. H., Garrett, T. A., Reynolds, C. M., Shaw, W. A., Moore, J. D., Smith, D. C., Jr., Ribeiro, A. A., Murphy, R. C., Ulevitch, R. J., Fearns, C., Reichart, D., Glass, C. K., *et al.* (2006). Kdo$_2$-lipid A of *Escherichia coli*, a defined endotoxin that activates macrophages via TLR-4. *J. Lipid Res.* **47,** 1097–1111.

Robinson, N. C. (1993). Functional binding of cardiolipin to cytochrome c oxidase. *J. Bioenerg. Biomembr.* **25,** 153–163.

Schlame, M., Brody, S., and Hostetler, K. Y. (1993). Mitochondrial cardiolipin in diverse eukaryotes. Comparison of biosynthetic reactions and molecular acyl species. *Eur. J. Biochem.* **212,** 727–735.

Schlame, M., and Greenberg, M. L. (1997). Cardiolipin synthase from yeast. *Biochim. Biophys. Acta* **1348,** 201–206.

Schlame, M., and Hostetler, K. Y. (1997). Cardiolipin synthase from mammalian mitochondria. *Biochim. Biophys. Acta* **1348,** 207–213.

Schlame, M., and Ren, M. (2006). Barth syndrome, a human disorder of cardiolipin metabolism. *FEBS Lett.* **580,** 5450–5455.

Schlame, M., Ren, M., Xu, Y., and Greenberg, M. L. (2005). Molecular symmetry in mitochondrial cardiolipins. *Chem. Physics Lipids* **138,** 38–49.

Schlame, M., Rua, D., and Greenberg, M. L. (2000). The biosynthesis and functional role of cardiolipin. *Prog. Lipid Res.* **39,** 257–288.

Sparagna, G. C., Johnson, C. A., McCune, S. A., Moore, R. L., and Murphy, R. C. (2005). Quantitation of cardiolipin molecular species in spontaneously hypertensive heart failure rats using electrospray ionization mass spectrometry. *J. Lipid Res.* **46,** 1196–1204.

Tamai, K. T., and Greenberg, M. L. (1990). Biochemical characterization and regulation of cardiolipin synthase in *Saccharomyces cerevisiae*. *Biochim. Biophys. Acta* **1046,** 214–222.

Taylor, W. A., and Hatch, G. M. (2003). Purification and characterization of monolysocardiolipin acyltransferase from pig liver mitochondria. *J. Biol. Chem.* **278,** 12716–12721.

Valianpour, F., Mitsakos, V., Schlemmer, D., Towbin, J. A., Taylor, J. M., Ekert, P. G., Thorburn, D. R., Munnich, A., Wanders, R. J., Barth, P. G., and Vaz, F. M. (2005). Monolysocardiolipins accumulate in Barth syndrome but do not lead to enhanced apoptosis. *J. Lipid Res.* **46,** 1182–1195.

Valianpour, F., Wanders, R. J., Barth, P. G., Overmars, H., and van Gennip, A. H. (2002). Quantitative and compositional study of cardiolipin in platelets by electrospray ionization mass spectrometry: Application for the identification of Barth syndrome patients. *Cli. Chem.* **48,** 1390–1397.

Vance, D. E. (2002). Phospholipid biosynthesis in eukaryotes. *In* "Biochemistry of Lipids, Lipoproteins and Membranes" (D. E. Vance and J. E. Vance, eds.), Vol. 36. Elsevier, New York. pp. 205–232.

Xu, Y., Malhotra, A., Ren, M., and Schlame, M. (2006). The enzymatic function of tafazzin. *J. Biol. Chem.* **281,** 39217–39224.

Author Index

Subject Index

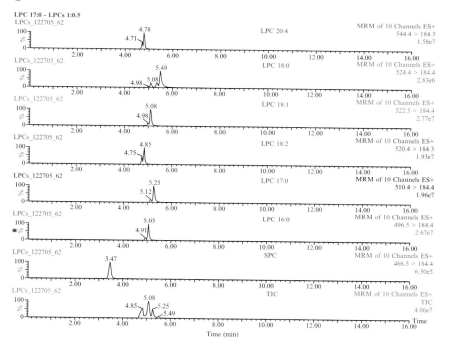

B

Mandi Murph *et al.*, Figure 1.2 Detection of lysophospholipids at negative (A) and positive (B) mode. (A) Mixture of standard LPS (18:1), LPI (18:0), LPA (16:0, 18:0, 18:1, 18:2 and 20:4), and S1P was analyzed and detected by negative MS/MS using sets of deprotonated molecular ion [M-H]⁻ and its daughter ion as described in section 4. (B) A mixture of LPC (16:0, 18:0, 18:1, 18:2, and 20:4) and SPC were detected by positive MS/MS. The sets of protonated molecular ions [M+H]⁺ and daughter ion m/z 184 [phosphorylcholine]⁺ were used. LPC 18:2 shows two peaks, which indicate the presence of both isomers—2-acyl type (fast-migrating peak) and 1-acyl type (slow-migrating peak).

A

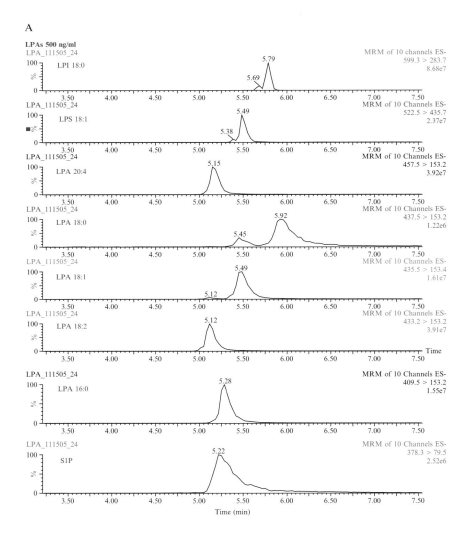

Figure 1.2 (*continued*)

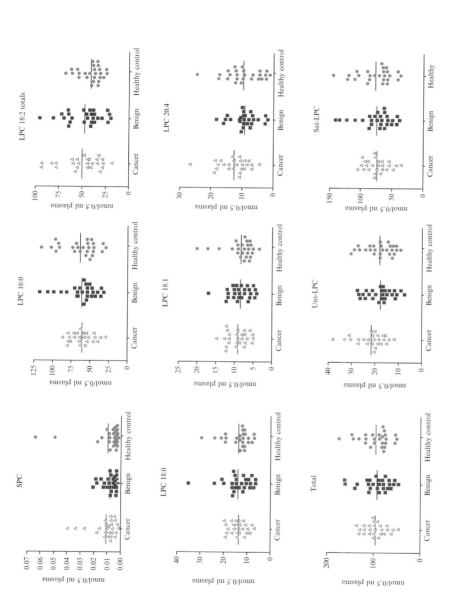

Mandi Murph *et al.*, Figure 1.4   Scatterplots of all lipids analyzed by LC/MS/MS. Plots show the comparison between lipids measured in cancer patients, women with benign tumors, and healthy controls. For more details, see section 4. Results were not statistically significant.

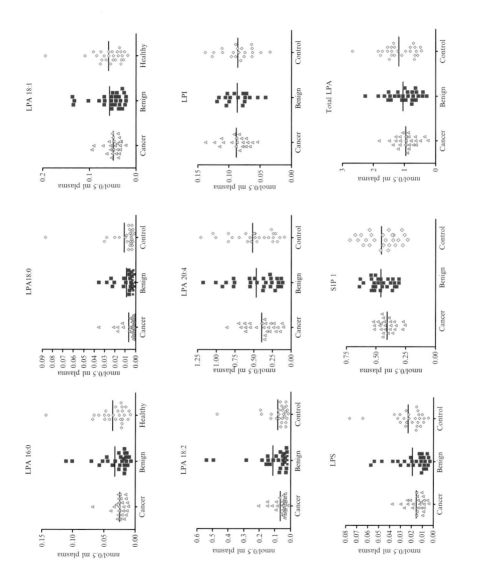

Figure 1.4 (*continued*)

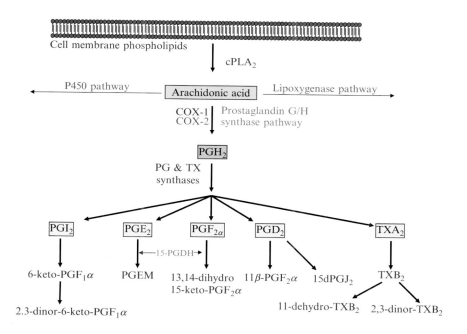

Dingzhi Wang and Raymond N. DuBois, Figure 2.1 Overview of PGHS (COX) pathway.

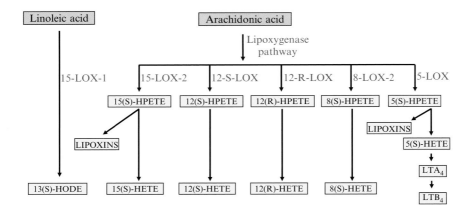

Dingzhi Wang and Raymond N. DuBois, Figure 2.2 Overview of lipoxygenase synthesis.

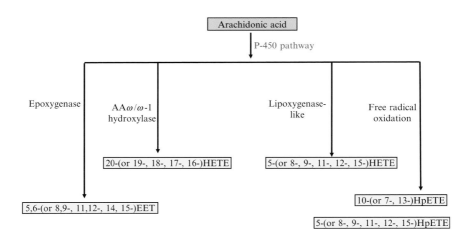

Dingzhi Wang and Raymond N. DuBois, Figure 2.3 Overview of P-450 synthesis.

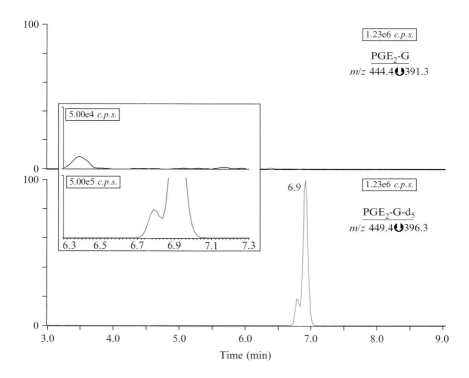

Philip J. Kingsley and Lawrence J. Marnett, Figure 5.6 Demonstration of isotopical purity of the internal standard PGE$_2$-G-d$_5$. The chromatogram is an injection of PGE$_2$-G-d$_5$ only. The upper transition is PGE$_2$-G, and the lower transition is PGE$_2$-G-d$_5$. The PGE$_2$-G channel is normalized to the intensity of PGE$_2$-G-d$_5$. The inset is a blow-up of the PGE$_2$-G-d$_5$ peak and corresponding time in the PGE$_2$-G transition. No PGE$_2$-G is seen.

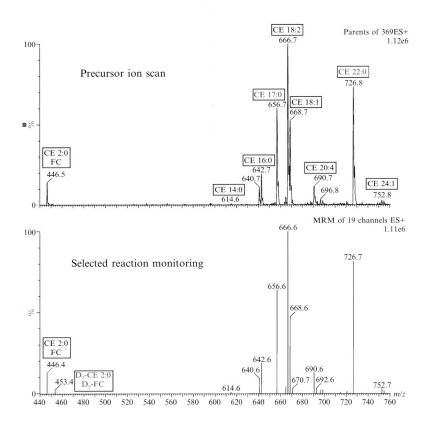

Dominik Schwudke *et al.*, Figure 10.5 PIS *m/z* 369 and SRM for CE and FC quantification. Free cholesterol was acetylated before analysis. The upper panel shows a PIS *m/z* 369.3 of a human LDL fraction in positive ion mode specific for CE. The lower panel shows the spectrum of the same sample reconstructed from SRM data. Note the internal standard for quantifying free cholesterol D_7-CE 2:0 was detected using different transition (*m/z* 453.4≤376.3). CE, cholesteryl ester; FC, free cholesterol.